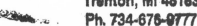
16

The Pursuit of Perfection

The Pursuit
of Perfection

THE PROMISE AND PERILS
OF MEDICAL ENHANCEMENT

Sheila M. Rothman
and David J. Rothman

Pantheon Books, New York

Copyright © 2003 by Sheila M. Rothman and David J. Rothman

All rights reserved under International and Pan-American Copyright
Conventions. Published in the United States by Pantheon Books,
a division of Random House, Inc., New York, and simultaneously
in Canada by Random House of Canada Limited, Toronto.

Pantheon Books and colophon are registered trademarks
of Random House, Inc.

Library of Congress Cataloging-in-Publication Data

Rothman, Sheila M.
The pursuit of perfection : the promise and perils of medical
enhancement / Sheila M. Rothman and David J. Rothman
p.   cm.
ISBN 0-679-43980-3
1. Medical innovations—Social aspects.   2. Social medicine.
3. Pharmaceutical industry.   4. Drugs—Research—Moral and ethical
aspects.   5. Medical ethics.   6. Perfection—Psychological aspects.
I. Rothman, David J.   II. Title.

RA418.5.M4R68 2003   362.1—dc21   2003046739

www.pantheonbooks.com

Book design by M. Kristen Bearse

Printed in the United States of America
First Edition
2   4   6   8   9   7   5   3   1

To our children

Contents

Introduction

THE PURSUIT OF BIOLOGICAL PERFECTION has an intriguing past, a dynamic present, and a formidable future, all testimony to an ongoing revolution in pharmacology and genetics. For well over one hundred years, scientists and physicians have designed and implemented enhancement technologies, drug companies have aggressively promoted and marketed them, popular culture has celebrated them, and individuals have zealously pursued them. The enterprise has never been able to free itself completely from quacks and nostrums, but it represents serious science and pervades medical practice. Its aim is not to cure a disease, to make a patient normal or remedy a deficit, but to make people better than well, to take them from standard levels of performance to peak performance.

The agenda is ambitious, even arrogant. In the coming decades, enhancements are likely to affect brain functioning, physical capacity, behavioral traits, and, with greatest potential impact on society, life span. Novel technologies may be able to improve memory and promote desired personality characteristics—such as risk-taking and sociability. Perfected bodies will be taller, leaner, stronger, quicker, and will require less sleep—no longer needing to go off-line eight hours a day. Once the genetics of aging are understood, longevity might increase not merely by ten or twenty years, which would be the result of curing today's leading causes of death, but by seventy to ninety years, with average life spans of 140 to 160 years.

Although many people are delighted by the prospect of such changes, the pursuit of perfection has also provoked considerable opposition, some critics uneasy with its mission, others steadfastly opposed. But however spirited the challenge, the historical record makes clear that the drive to enhancement is so deeply entrenched in our science and culture that neither religious nor social objections will be able to withstand, divert, or regulate it. At times, a specific technology may be temporarily slowed by popular objections and government regulation—DNA research in the early 1970s is one case in point, and stem cell research today is another. But sooner rather than later the research proceeds. Science exerts too powerful a hold on its practitioners and on the public to be contained.

Which brings us to the question at the heart of this book: What will the all-out pursuit of biological perfection mean to us as individuals and as a society? There has been much armchair speculation about its consequences, but we will take a different tack, drawing on the rich and intriguing historical record of enhancement. By looking back, particularly on the history of such hormones as estrogen and testosterone, and looking now at plastic surgery and growth hormone, we will be better able to understand the promise and perils of novel genetic enhancements, framing the real and specific challenges that lie ahead.

I

To generations of researchers, the idea of improving on nature by stripping her of her secrets has been an article of faith. From at least the mid-nineteenth century onward, and particularly in the aftermath of Darwin, the overriding mission of biology has been to refashion nature. From the celebrated French physiologist Claude Bernard in the 1850s, to such Nobel Prize–winning geneticists as Hermann Muller in the 1930s and Joshua Lederberg in the 1950s, researchers have been unrelenting in their efforts to penetrate nature so as to improve it. Muller expressed the boundless ambitions of biology when predicting that the discipline would "reach down into the secret places of the great universe" to produce "an increasingly sublime creation—a being beside which the mythical divinities of the past will seem more and more ridiculous."[1] Half a century later, Lederberg echoed this same value. "Many people find it difficult to accommodate the reality that Nature is far from benign. Those who are horrified at any tinge of our 'tampering with natural evolution' need to be reminded that this has been intrinsic to human nature since Prometheus." Nature should not be admired from a distance but altered for the better. "Contemporary man," Lederberg concluded, is not a natural species but a "manmade species."[2]

What science creates medicine rapidly dispenses. Ours is a culture in which physicians enjoy great discretion in defining what constitutes a disease and what complaints warrant their interventions. The catalogue of conditions that belong in a doctor's office has expanded dramatically over the past one hundred years, to the extent that twentieth-century medicine is comfortable in taking human unhappiness as its province. Should unhappiness reflect a dissatisfaction with short stature or body fat or signs of aging, then medicine is right to apply its knowledge of hormones or its surgical techniques in order to relieve it. If in doing so, physicians cross the admittedly flimsy line between cure and enhancement, that is of little concern.

Not only science and medicine but the drug industry is poised to promote enhancements. Pharmaceutical houses' quest for profits ensures that new technologies will be produced, advertised, and distributed widely. The close ties that drug companies have forged with organized medicine and individual practitioners feed this effort. Salesmen, or detail men as they are known, regularly visit physicians' offices, with journal reprints and gifts in hand, encouraging doctors to respond to patients' complaints of unhappiness, or fantasies of perfection, by prescribing their company's products.

Medicine's enthusiasm for enhancement at once stimulates and is reinforced by a culture that prizes individual perfection and peak performance. Let an enhancement technology appear to deliver a contoured body free of the visible signs of aging and the media will cover it extensively; they will be certain to interview the innovator as well as the first delighted patients. Drug companies will advertise it directly to consumers and hospitals will spread the word that they have the technology and use it more often and more successfully than their competitors. Consumers will then try it for any number of reasons: to obtain or retain a job, to win a promotion, to please themselves or to please others, including spouses or potential spouses, or to compete with friends and neighbors.

The system, however, is out of balance, for no part of it has a stake in emphasizing or even communicating the dangers that are almost certain to accompany the innovation. Most commentators on enhancement presume that the technologies will arrive full-blown and effective, able to deliver immediately on their promise. But the record strongly suggests that technologies will emerge slowly and haltingly, some delivering benefits, others inflicting serious harm. Consumers will be compelled, personally and collectively, to make a series of exquisite choices, with very little data to guide them. In fact, the choices about potential enhancements will be far more complex than choices about would-be cures. Illness carries its own risks, which have to be balanced against the likelihood that a new treatment will prove effective. But with enhancement, healthy adults will have to calculate how much risk they are willing to accept in order to try to optimize a trait. Is it wise to undergo an intervention that promises to dramatically increase life span and disregard the risk that it might cause fatal disease and shorten life span?

To further complicate decision making, the most influential voices coming from science, medicine, culture, and commerce are likely to emphasize potential benefits and discount the dangers. The researchers who devise the breakthroughs will pay scant attention to its clinical uses, assuming it is the doctors' responsibility, not theirs. Many physicians will be ready to dispense it, assuring their patients that in their hands all will go well. Medicine's track record has never been very good in communicating the risks of treatment for major illnesses, and there is little reason to believe it will be more proficient with enhancements. Nor is the performance of the drug companies impressive. They have failed in too many instances to report adverse outcomes and they are not likely to be the first to sound alarms. Thus, consumers will be on their own, forced to sort out for themselves the possible side effects, which are not going to be trivial. Whatever the enthusiasm of their inventors and promoters, enhancement technologies will not present a win-win situation. There will be victories and defeats, and only by understanding the complicated and compromised positions of science, medicine, and commerce will it be possible even to start to calculate the odds.

II

The prospect of potent and effective enhancements has generated substantial resistance, often from members of the loosely knit bioethics community. They offer different justifications for their hostility but join together in the hope of subverting the enterprise. Some are persuaded that such concepts as the "natural" and the "unnatural" should constrain science and medicine, insisting that these terms provide a firm conceptual basis for leaving the human condition unmodified, no matter what technology might be able to accomplish. Others are convinced that enhancement is at its core frivolous, looking to satisfy vanity and self-indulgence. Enhancement may even threaten the moral fiber of the citizenry, replacing a commitment to earn attributes and rewards with genetic or pharmacological shortcuts. Some fear that the technologies will widen the gap between social classes, transforming economic differences into biological ones. The values of equity and fairness will be brushed aside in the rush to enhancement. Still others believe that the grim history of eugenics casts its shadow on the entirety of the enterprise. A world that remembers the evils perpetrated by Nazi Germany cannot dare encourage scientific pursuits that may turn into tools of repression.

Although these arguments are not without merit and warrant consideration, the negative positions do not reckon with either the complicated dynamics that are certain to propel enhancements forward or the more immediate and real dangers. The most useful starting point for analyzing them is with the meaning of *cure* and *enhancement*, the two terms that run through the debate. At first glance, differentiating between the two seems relatively simple. Cure seeks to restore health by remedying a disease, by returning tissue or body chemistry to a normal range as set by the general population or patients of the same age. By contrast, enhancement looks beyond the normal, aiming to put the individual at the far end of the curve, or better yet, off the chart.

But it turns out to be extraordinarily difficult, really impossible, to distinguish consistently between cure and enhancement, to compose a chart with interventions that aim to cure on one side of the page, and enhancements on the other. Cure itself is a highly ambiguous concept that requires precise definitions for such amorphous terms as health, disease, normal and abnormal, definitions that cannot be provided with any consistency or confidence. The most obvious examples of how treacherous it is to invoke such terms come from the field of psychiatry. In the 1970s, it attempted, without success, to define where mental disease left off and alternative but normal lifestyles began. In the 1980s, it debated whether homosexuality was a disease (the original position), or a life choice (which it became). In the 1990s, it pondered whether a number of behaviors, from short attention span to frequent gazing into the mirror, were or were not diseases.

Medicine confronts the same dilemmas. Is short stature a disease? Is aging? Is muscular weakness in a seventy-five-year-old? Is a crooked nose or bulging thighs? Do these conditions become diseases if they provoke personal unhappiness? Moreover, medicine, like psychiatry, has had great difficulty in defining the normal. Endocrinology and gynecology have struggled to establish the characteristics of a "normal" menstruation—when it should start, what degree of premenstrual tension is to be expected, and when menopause should begin. But because variations among women are extensive, and because terms such as *tension* do not lend themselves to precise definition, one physician's finding of normal is another's abnormal—which means that a justification for intervening is always close at hand. So too, the criteria for what is a normal test result change frequently. What was considered a "normal" blood pressure or cholesterol level in the 1980s is now considered a risk factor and grounds for prescribing different types of drugs. Medicine has also changed its mind on what constitutes normal levels of testosterone in older men, not because of new findings or diagnostic innovations but, as we shall see, because drug companies made testosterone more readily available in potent forms.

The futility of trying to separate cure from enhancement is also apparent in the way that science conducts research and medicine translates its findings into clinical practice. In 1997, the National Institutes of Health (NIH) convened a meeting on the regulation of genetic enhancement to consider banning funding for research on enhancement.

The policy was never enacted because of an inability to differentiate one type of research activity from the other.[3] As we shall repeatedly discover, investigations that begin with an intent to treat often lay the groundwork for an enhancement. Thus, the task of isolating and synthesizing growth hormone started as part of a research effort to treat short children who were growth hormone deficient. However, once the substance was in ample supply, it was administered not only to children with measurable deficiencies but also to those who had the "disease" of short stature. Next it was dispensed to increase muscle mass in elderly men whose physical weaknesses could be diagnosed either as a normal part of aging or as a disease. So suppose an investigator applies for a grant to study growth hormone. Should NIH fund it because hormone deficiencies are a disease, or summarily reject it because the research might help produce an enhancement? Rather than attempt to resolve such tortuous questions, the NIH retreated and issued no policy guidelines to distinguish cure from enhancement.

The history we will trace is filled with examples of cures turning into enhancements. Hormone replacement therapy began as a cure for the acute symptoms of hot flashes and anxiety that sometimes accompanied the passage to menopause; later it became an intervention to keep skin supple and improve memory in postmenopausal women. Plastic surgery was developed to rehabilitate soldiers injured and maimed in World War I; the same techniques were later applied to the lined face and the small breast. Testosterone was used first as a means to treat sexual underdevelopment in young men and then was dispensed in an effort to reverse the "normal" course of aging. Indeed, it is most likely that the knowledge necessary to design future enhancements will emerge from research now being conducted on leading diseases. Advances in the treatment of Alzheimer's disease may someday produce memory enhancement for the normal population. Progress in understanding the biology of cellular division in cancer cells may uncover techniques for creating cell immortality that may double average life span.

Taken together, the difficulties of drawing a hard line between cure and enhancement undercut some of the objections to enhancement. If the two cannot be neatly distinguished, then judging results of one intervention as natural and the other as unnatural, one as serious and the other as trivial, has little meaning. Once you accept the moral legitimacy of an effort to cure, it becomes very difficult to build an argument for the moral illegitimacy of an effort to enhance.

III

For many critics, the core objection to the pursuit of perfection is that the results will be "unnatural" and diminish our essential humanity. They decry the hubris of science and its overarching ambitions. "Enchanted and enslaved by the glamour of technology," the conservative bioethicist Leon Kass insists, "we have lost our awe and wonder before the deep mysteries of nature."[4] Francis Fukuyama addresses the "consequences of the biotechnology revolution" in terms of *Our Posthuman Future*.[5] Others contend that "in order to stop the dehumanization of man, we may have to forgo some research. We may have to say no to some research before it begins."[6] Although an appeal to the natural has a long and venerable tradition—think of the elaborate formulations of natural law and natural rights—when put into the context of a specific human trait or condition, the yardstick falls short on several counts. First, there is no consensus in biological terms as to what defines the critical elements of our humanity, what constitutes "the natural" that should be respected, or the "unnatural" that should be avoided. By what grounds is it "natural" to extract a kidney or a heart from a person who has just died and transplant it into a living being? Is there not something fundamentally unnatural about living with some one else's heart? Why would it be unnatural to extend longevity to 140 years? The average life span almost doubled between 1800 and the present (with no objections voiced). Why not allow it to double again? Would it truly be unnatural to sleep fewer hours or have a better memory? Examples could be compounded but the point is clear: what "is" now does not seem a useful standard for deciding what "might be."

If the appeal to nature is not robust enough to restrain enhancement technologies, neither is the charge of frivolity. What appears trivial to one person is of crucial importance to another. Take the use of growth hormone in short but not hormone-deficient children. Critics insist that physicians ought not to indulge parental fantasies of children growing tall enough to play professional basketball. But in fact, most parents and children are grappling with the stigma and emotional pain of being the "shrimp" and serving as class scapegoat or mascot. The driving force is a profound unhappiness, not a future in the NBA. So too, some people may find it admirable to shape the body through diet or exercise, but others may prefer plastic surgeries. Is this an indulgence, a convenience, an exercise in narcissism, or a shrewd career move? Each of us may have personal values and opinions on the question, but that hardly justifies imposing them on others.

The more difficult challenge to enhancement is the historical one: Should the appalling abuses in twentieth-century eugenics restrain the pursuit of genetic enhancements? The Nazi experience cannot be forgotten. Its concepts of race improvement led to barbaric practices, the forced killing of so-called mental defectives, terminally ill persons, ethnic minorities, homosexuals, and then, in the greatest numbers, Jews. But this horrific record is only marginally relevant to our own deliberations. It reflects the pivotal role that the German state played in coercing weak and vulnerable citizens. It was eugenics imposed and dictated from above. By contrast, what we are witnessing now is enhancement embraced, pursued from below. Its goals are not to improve a nation's stock by altering the composition of its population (as in eugenics), but to elevate individual human capacities. The search is for methods to optimize individual performance not to secure group homogeneity.

To be sure, an unanticipated consequence of enhancement technology might be to encourage its own special form of coercion. Were some number of individuals to enhance their memories or physical size, others might feel compelled to emulate them, fearful of finding themselves at a marked disadvantage. One can imagine, as well, a government that would define an enhancement as so fundamental to the public health that it mandates the use of the procedure, a futuristic twist on the tradition of compulsory vaccination. But however portentous such concerns, they fall well outside the historical reference points for eugenics. The heightened individualism in our society is so consistent with enhancement that such technologies are far more likely to be sought after by citizens than imposed by a government.

Should genetic enhancement be restricted because of the possibility of social injustice? Is it proper to restrain the technologies for fear that they will produce "a group of privileged individuals whose position in society will be virtually unassailable," or, as the biologist Lee Silver fears, create "segregated social worlds," elevating "gene-enriched individuals" over the "natural"?[7] Such an outcome is not unimaginable but we must ask why this technology should be singled out when no other one is so treated. Given the persistent inequities in American health care delivery and in American society more generally, there is almost no medical advance that can promise to pass the test of fairness. Moreover, it is not fanciful to imagine that genetic enhancements might cross class lines. To the degree that technologies improve physical capacity and mental acuity, public and private investments in their distribution might be forthcoming. None of this is certain, but criteria of social justice are most useful when invoked not as barriers to innovation but as rallying cries for ameliorative action.[8]

IV

In the pursuit of perfection, it is women who most often occupy center stage. They are the primary, although not exclusive, objects and consumers of medical enhancements. Our culture encourages women, much more so than men, to focus on their bodies, to carry on a highly charged dialogue with stereotypic images. The long and tangled history of estrogen and the rise and fall of hormone replacement therapy is one prime example. Investigators in biology, endocrinology, and gynecology—long fascinated by the menstrual cycles that women move through over the course of a month and over the course of a lifetime—were confident that if only they could identify the underlying physiological causes and isolate and administer the substances that triggered the changes, they would be able to cure such conditions as sexual underdevelopment and infertility, and optimize women's energy, appearance, and sexuality as they aged. Even without solid data on benefits and risks, gynecologists made estrogen the centerpiece of their practice, and drug companies supported research to help develop new markets and new uses for this increasingly prescribed drug.

Plastic surgery also provides a means to satisfy women's desire to reshape their bodies, even as the competition between surgeons and dermatologists for control of the lucrative liposuction market exposed women to ever increasing risk. Meanwhile, rival feminist camps debate whether such enhancements are demeaning to or empowering of women, a diversity of opinions that gives abundant space and legitimacy to a variety of enhancements and helps explain why some women will take risks, even inordinate risks, to achieve a benefit that others will consider trivial or demeaning.

V

Another hormone that occupies a special place in the history of enhancement, and moves our story from women to men, is testosterone. Like estrogen, it was the object of extensive laboratory research that disregarded conventional standards of the natural. As soon as testosterone was synthesized, physicians eagerly prescribed it, letting logic and metaphor substitute for data. Like an automobile running low on gas after many miles, the aging male was running low on testosterone. Fill his tank, bring his testosterone levels to those of a twenty-year-old, and he will become as a twenty-year-old again, with all his muscles, energy, mental capacities, and sexual capacities.

Physicians and drug companies tried their best to create a market for testosterone, hoping that the lure of such exceptional benefits would overcome men's general reluctance to go to doctors and the fact that they did not experience a readily identifiable menopause. Despite little evidence of benefits—and great possible risks—anti-aging clinics around the country continue to dispense testosterone to older men, promising to reinvigorate their aging bodies.

Even children have been the objects of enhancement, providing a compelling example of how inseparable cure and enhancement are. No sooner was human growth hormone synthesized than it was administered widely not only to children who were hormone deficient but also to those who were of short stature, but did not lack the hormone. What little data there was strongly suggested that the hormone could not increase their height, but doctors prescribed it anyway—some lured by the marketing schemes of drug manufacturers, and some unable to resist parents' pleas to do anything to help their short children grow taller. Moving forward, human growth hormone has joined testosterone as a staple in anti-aging clinics, promising regenerative effects, and is widely dispensed despite dubious benefits and almost certain risks.

What is the level of caution that should be exercised when it comes to enhancement? This question becomes all the more relevant as new genetic interventions emerge that promise to enhance our brains, bodies, and behaviors. The possibilities are awesome. These technologies may double or triple powers of recall, increase muscular strength and endurance, and not only extend the years of physical and mental competence but revolutionize longevity. Yet, the historical record makes apparent that the new procedures will come with risks as well as benefits. One can already foresee the future narrative. A purported enhancement technology explodes into prominence. The front-page story or lead article in a prestigious medical journal announces a fantastic breakthrough, focusing not on the small size of the sample or the precise limits of what was actually accomplished, but on the grandiosity of the promise. A year or two later, follow-up studies appear, with far more ambiguous results. Still later, more studies emerge, refuting the entirety of the claims for benefits and identifying serious side effects, that may include death. In the end, individuals will have to make their own calculations, without much guidance. There will be winners and losers, but no one will be able to predict whether they will be the risk-takers or the risk-aversive.

VI

As we researched and wrote this book, a few friends and colleagues asked us a question that we had not anticipated: Would we do research firsthand? Would we test enhancements on ourselves, see what worked and did not work, and report our findings? The question, in fact, has a very solid historical foundation. Enhancements were often first tested by their designers; their own needs inspired the research and their own responses served as data. The most famous example is the physiologist Charles

Édouard Brown-Séquard, who in 1889, at the age of seventy-two, in-jected himself with a mixture of testicular substances and declared him-self reinvigorated.

We did consider the idea of testing some enhancements on ourselves, but before we could even begin to ponder potential risks and benefits, we recognized the impossibility of such a venture. There are too many would-be enhancers now available. Every body part has its own plas-tic surgeon and cosmetic formula. The drugs purporting enhance-ment effects include estrogen, testosterone, growth hormone, Viagra, Prozac, and DHEA along with vitamin E, vitamin C, and gingko, to name only a few. Juice stores advertise "immune boosters" and "femi-nine boosters." Anti-aging clinics offer all these products and many more. One such clinic would be happy to measure our levels of beta-carotene, hydroperoxides, tocopherol, melatonin, trace metals (tin, zinc, mercury, nickel), and oxidative stress, and then supplement the levels as it believed necessary.

This state of affairs simplified our decision. It became quickly appar-ent to us that the field of enhancement needs outside evaluators far more than it needs human subjects. Our contribution is best made by analyz-ing the record, not becoming part of it. We are hopeful that the pages that follow justify our approach.

The Pursuit of Perfection

ONE

Penetrating Nature

FOR THE PAST 150 YEARS, the biological sciences have been at war with nature, determined to penetrate its secrets in order to perfect it. Investigators are eager to intervene, tamper, modify, and revise the forms and substance of animal life and human life. They may share a grudging respect or even wholesale admiration for the complexity of a biological system or a particular organism, but their readiness to improve on it is not inhibited. The proposition that the natural represents what can be or should be is altogether alien. No matter how intricate the existing design, it may still be enhanced, even if the result might be unusual or unimagined. Biology has no fixed boundaries, only opportunities.

When confronting resistance from secular or religious spokesmen, some biologists attempt to deflect criticism by conflating scientific methods with scientific goals. Since their methods are objective and neutral, their findings should be considered objective and neutral. Others emphasize the tangible benefits of the research, offering many examples of better and longer living through biology. To be sure, critics challenge their contentions, rejecting the idea of scientific neutrality or the premise that a longer life is a better life. Although their negativity may appear to represent a post-atomic age or post-genomic era response to science, its roots are much deeper. In 1818, Mary Shelley was asking whether Dr. Frankenstein ought to be plundering the graveyards of body parts in order to try to create a living being, and in 1896, H. G. Wells wondered whether Dr. Moreau ought to be transplanting body parts between animals so as to construct a more perfect beast. But however persistent the hostility, modern biology has not altered its fundamental approach to nature.

Its ambitious, even combative, attitude was articulated with particular brilliance and confidence by the pioneering figure in the field, Claude Bernard. Chair of the Department of Medicine at the Collège de France and a founder of the discipline of physiology, Bernard unabashedly defined nature as the enemy to be conquered. His 1856 book, *An Introduction to the Study of Experimental Medicine,* set forth a history of science and a method for future biological research.[1] His predecessors, Bernard contended, had done little more than use their senses to observe biological processes. Early medical practitioners, for example, felt a patient's pulse, looked to see if the face was flushed, examined the urine, and listened to the chest. If they found a rapid pulse, they tried one remedy; if the urine was cloudy, they tried another. But they never actively intervened to understand the mechanisms responsible for one or another physical condition. In more modern times, biologists and physicians went one step further, observing nature through careful groupings of facts and testing of hypotheses. They treated patients with a drug, analyzed the results, and tried to evaluate its efficacy. But this approach, too, had only limited value, for the physician remained a mere observer of symptoms and outcomes.

Bernard championed a third and very different kind of biology, the "active observation" of nature. To characterize it, he quoted an aphorism from his French naturalist colleague Georges Cuvier: "The experimenter questions [nature] and forces her to unveil herself."[2] As Bernard went on to explain: "Experimenters must be able to touch the body on which they act, whether by destroying it or by altering it, so as to learn the part which it plays in the phenomena of nature. . . . It is on this very possibility of acting, or not acting, on a body that the distinction will exclusively rest between sciences called sciences of observation and sciences called experimental."[3]

Bernard's language demonstrates how aggressive experimental science sought to be. The investigator "touches" the body, not gently or respectfully but in ways that alter or even destroy it in order to learn about its functioning. Like a sexual predator, he forces nature to "unveil herself " so he may penetrate her mysteries and fathom her secrets.[4] For Bernard, this very aggression differentiated an older and weaker science of observation from a newer and wiser science of experimentation.

This approach, Bernard predicted, would empower biology to transform nature. "Man becomes an inventor of phenomena, a real foreman of creation. . . . We cannot set limits to the power that he may gain over nature through future progress."[5] Once a scientist made phenomena appear "under conditions of which he is the master," that is, in his laboratory, he would be able to "dominate nature . . . to conquer living nature, act upon vital phenomena and regulate and modify them."[6] Bernard was acutely aware of a cultural resistance to the idea of scientist as foreman of creation. But rather than compromise his position, he urged colleagues to ignore popular opinion. Since "it is impossible for men, judging facts by such different ideas, ever to agree . . . , a man of science should attend only to the opinion of men of science who understand him."[7]

Bernard's vision for biology won over the discipline not only because it promised to unleash the power of science but because it fit so well with the new and powerful framework that Charles Darwin provided to order the natural world. Biological change, as Darwin explained, was inevitable and seemingly ungovernable. *On the Origin of Species* made clear that the static, hierarchical, and fixed vision of nature that marked earlier thinking had to be replaced, really swept away, by a far more fluid vision consistent with the dynamics of evolution. Darwin, as his biographer Janet Browne observes, was inviting people to "believe in a world run by irregular, unpredictable contingencies."[8] A particular species appeared at one stage, adapted itself at another, and become extinct at still another. The mechanisms responsible for these changes were "natural," that is, the result of a non-human (and non-divine) process of selection that no one could control. "The fitness of an individual organism to its environment," as another Darwin scholar, Gillian Beer, has noted, "increases the chance of survival." But the environment is "a matrix of possibilities, the outcome of multiple interactions . . . prone to unforeseeable and uncontrollable changes." Precisely because the "everyday does not last forever," either on an individual or collective basis, "will and endeavor must always be insufficient. They can never control all the multiple energies of life." The variegated and all-powerful forces of nature would overwhelm the best efforts of any single individual or organism.[9]

But rather than intimidate biology, Darwin actually liberated and emboldened it. His impact is well exemplified in the ambitions of Jacques Loeb. Now more famous in his fictional than in his real-life guise—he was the model for Max Gottlieb in Sinclair Lewis's *Arrowsmith*—Loeb was a well-honored biologist in the opening decades of the twentieth century. More cogently than others, he staked out a position that his biographer, Philip Pauley, aptly labels the "engineering standpoint" in biology, and helped embed it in American biology.

Loeb made Darwin his starting point for promoting science as a form of action, encouraging biologists to manipulate nature. Giving the "natural" a privileged position within biology made little sense because the natural was nothing more than the result of a mutation that provided one animal with an advantage in survival over another. By definition, this mutation was accidental in origin even as it became the standard for the species; change was driven by chance. Why, then, should biologists give deference to the outcome of chance? Why not encourage science itself to create the variation? Surely human forethought and design were no less desirable than accident in altering nature.

Accordingly, Loeb proposed that biology "gain a deeper and more certain insight into the possibilities for the transformation of the species beyond that which we have at present." The challenge was open-ended, carrying limitless possibilities that Loeb readily described. A biology that was free to innovate might enable investigators to transform human life to the point of preventing "the wasting of the body in old age." They might even discover the means of transforming dead matter into living matter (Frankenstein's ambition realized). The decisive element was to liberate the laboratory from restraints, ideological and doctrinal, so that it could make "the attempt of controlling at will the life phenomena of animals and of bringing about effects which cannot be expected in nature." Here was the foundation for a biology of enhancement.[10]

Yet another consideration drove this ambition forward—the exuberance that came with biology's initial successes in improving upon nature. By the opening decades of the twentieth century, scientists were exhilarated by the knowledge and power that they had already achieved, and with almost childish enthusiasm, anticipated future achievements, provided, of course, that no one blocked their way. Perhaps the most extravagant expression of this outlook came from one of England's most versatile scientists, as comfortable in chemistry as he was in biology, J. B. S. Haldane. To appreciate how confident investigators were about remaking nature and assuming the role of biological and social engineers, the best place to begin may well be Haldane's short and best-selling 1924 book, *Daedalus, or Science and the Future.*[11]

Haldane's premise was that science must enjoy complete freedom to pursue new knowledge. Indeed, he was remarkably optimistic that this freedom would be realized for two shrewd and interrelated reasons. The essential spirit of capitalism, which rewarded all types of entrepreneurial activity, was now joined to a pervasive "competitive nationalism," which led countries everywhere to place a premium on scientific advancement. These two forces combined guaranteed that few impediments would block "the advantages accruing from scientific research." With remarkable prescience, Haldane recognized that the drive for markets together with patriotic zeal would ensure that science would be both handsomely supported and left independent.[12]

To Haldane as well as to most other biologists, religious objections to tampering with the natural or secular sensibilities that privileged the natural carried no weight. As Haldane dismissively put it: "There is no great invention, from fire to flying, that has not been hailed as an insult to some God."[13] Unrestricted by superstition or ignorance science had already made great progress. Haldane predicted further innovations in travel and communications that would enable "any two persons on earth . . . to be completely present to one another in not more than $\frac{1}{24}$ of a second." (The Internet foreseen?) Biology and medicine had already extended the life span of individuals and it was not fanciful to contemplate "the abolition of disease."[14] Not only cures but also enhancements were certain to follow. Haldane expected medicine to realize "the direct improvement of the individual," especially through endocrinology. "As our knowledge of this subject increases," he predicted, "we may be able, for example, to control our passions . . . to stimulate our imagination . . . [and] to deal with perverted instincts by physiology rather than prison."[15] Forecasting the synthesis of the female hormone estrogen, he speculated that as biology better understood the chemical substances produced by the ovaries and then isolated and duplicated them, "we shall be able to prolong a woman's youth, and allow her to age gradually as the average man."[16]

Along with other biologists across the political spectrum, Haldane linked the science of eugenics to social advances. New laboratory techniques would create what he called "ectogenic children," born from embryos that had been fertilized and nurtured outside the womb. (It was Haldane's ectogenic fantasy that inspired Aldous Huxley to write *Brave New World* and popularize the idea of test-tube babies.) Predicting that by 1968 France would be using test tubes to produce 60,000 specially gifted children a year, that the pope would condemn the procedure, and that many people would miss the "old family life," he blithely concluded that "the effects of selection have more than counterbalanced these evils." The "output of first-class music" and "decreased convictions for theft" were two of his examples. "Had it not been for ectogenesis," he declared, "there can be little doubt that civilization would have collapsed within a measurable time owing to the greater fertility of the less desirable members of the population in almost all countries."[17]

In all, Haldane was certain that "scientific knowledge is going to revolutionize human life," asserting the proposition with a vigor that seems not merely excessive but naive to a post-Nazi, post-Hiroshima generation. To be sure, he conceded that "man armed with science is like a baby with a box of matches." But scientific progress would allow man to conquer space, conquer time, overcome "the dark and evil elements in his own soul," and ultimately refashion "his own body and those of other living beings."[18]

These fantasies were shared by others, none more inventive than H. G. Wells, the British science-fiction writer and socialist agitator. In 1895, Wells published a brief essay, "The Limits of Individual Plasticity," and not only the title but also the contents have a distinct postmodern ring. Invoking Darwin, Wells saw no reason to "give mere subservience to natural selection, or to respect inherited traits as though a living thing is . . . nothing more than the complete realization of its birth possibilities." Rather, "we overlook only too often the fact that a living being may also be regarded as raw material, as something plastic, something that may be shaped and altered . . . and the organism as a whole developed far beyond its apparent possibilities." Through science, "a living thing might be taken in hand and so moulded and modified that at best it would retain scarcely anything of its inherent form and disposition." Biology would so extensively recast physical and mental capacities as to "justify our regarding the result as a new variety of being."[19] Wells did not specify just how the discipline would fulfill this grand challenge, but he did speculate about a future rich with organ transplantation and plastic surgery accompanied by medicine's ability to modify "the chemical rhythm of the creature . . . and methods of growth." He did not "believe that the last word, or anything near it, of individual modification has been reached."[20] We could look forward to living creatures made over "into the most amazing forms."[21]

Biology's impatience with nature and its readiness to improve on it also received preeminent expression in the life's work of Nobel Prize winner Hermann Muller. In the 1930s, Muller demonstrated that the application of X rays to fruit flies created mutations. Not all of these induced changes, of course, were favorable to the survival of the species, but his X-ray technique exemplified the ability of science to alter the genetic composition of living things, and, in effect, to shape the course of evolution. Through its ability to create amazing forms—now in fruit flies but eventually, perhaps, in humans—science was confirming that purposeful activity by man could replace the accidents of nature.[22]

Muller was especially articulate about the propriety and benefits of such strategies. The very title of his 1959 essay, "The Guidance of Human Evolution," in which he reviewed his research career, expressed the essence of his position: "Natural selection is too opportunistic and shortsighted to be trusted to give an advantageous long-term result for any single group of organisms." Among the various species, mankind had been fortunate, so far, to survive and not vanish like so many others, but Muller worried that in time humans might suffer such a fate. He also feared that medical advances joined to an expanded sense of social responsibility might corrupt future genetic stock. "It is regarded as ethical," he lamented, "to employ every artificial aid to enable an individual

to reproduce . . . even when his reproduction would be likely to perpet-
uate the genetic condition that had occasioned the given difficulty." (His
case in point was the medical advances that allowed diabetic women to
bear children, thereby increasing the number of diabetics in the society.)
"From now on," he concluded, "evolution is what we make it." It must
become "a conscious process . . . [so] it can proceed at a pace far outdis-
tancing that achieved by trial and error." If man "sees some grand
process like evolution, and that it would be at all possible for him to be in
on that game, he would irreverently have to have his whack at that
too."[23]

   To be sure, scientists had their differences on how best to re-engineer
nature. Some insisted that the chief engine of change should be a coercive and negative eugenics. Charles Davenport, zoologist and eugeni-
cist, and his followers at Long Island's Cold Spring Harbor took this
position. As Daniel Kevles has explained, they were prepared to violate
individual rights to promote their own definition of the good of the
race. The "most progressive revolution in history" could be realized
if only "human matings could be placed upon the same high plane as
that of horse breeding."[24] But most others adopted a less heavy-handed
approach. Biology could promote progress without resorting to coercion. Haldane, for example, vigorously opposed state-sponsored efforts
to segregate and sterilize the mentally disabled: "Many of the deeds
done in America in the name of eugenics are about as much justified by
science as were the proceedings of the inquisition by the gospels."[25] In
sum, scientific engineering was embraced by both political liberals and
radical conservatives. The appeal of having biology perfect nature
crossed ideological boundaries.

I

The first tangible evidence that science and medicine might be able to
accomplish this grand mission came in the mid-nineteenth century from
the new field of physiology and the specialty that it spawned in the
opening decades of the twentieth, endocrinology. Claude Bernard, himself a physiologist, was among the first to recognize the awesome bio-

logical powers of what he called the "internal secretions of the ductless glands," what we know as hormones. From Bernard onward, physiologists played a crucial role in explaining how these internal secretions help to preserve health, to determine sexual characteristics, to facilitate reproduction, and to influence physical size and cognitive functioning.

Physiology took hold in French and German laboratories in the 1850s, commanded the attention of English scientists in the 1870s and 1880s, and exerted a critical influence on American science and medicine in the 1890s.[26] The field defined itself in contradistinction to anatomy, the oldest of medical disciplines, going back at least to the Roman physician Galen, which looked at the physical structure of the body. It was an essentially descriptive exercise, laying out, in Bernard's negative terms, the "passive mechanical arrangements of the various organs."[27] Although biology had to understand these arrangements, anatomy by itself was incapable of analyzing the contributions of particular organs to sustaining life processes. "Dead anatomy," Bernard asserted, "teaches nothing. . . . To know something about the functions of life, you must study them in the living. . . . It is in the study of these inner organic conditions that direct and true explanations are to be found for the phenomena of the life, health, sickness and death of the organism."[28] In other words, anatomy was a static discipline, and physiology a fluid one. Only physiology could imagine a body in which particular organs and tissues substituted for one another, or a body in which some organs and tissues either increased or decreased their activity. Anatomy described what was. Physiology had the power to analyze what is and to create what might be.

All of physiology's methods depended on vivisection. As Bernard unabashedly explained, "To analyze the phenomena of life, we must necessarily penetrate into living organisms, with the help of the methods of vivisection."[29] Recognizing that he was on controversial grounds— ethical objections to animal research were already being raised— Bernard vigorously defended the practice. Distinguishing carefully between human and animal research, he insisted that investigators never conduct an experiment on man "which might be harmful to him to any extent, even though the result might be highly advantageous to science, i.e., to the health of others."[30] This very principle, however, made ani-

mal research more vital. "It is essentially moral to make experiments on
an animal, even though painful and dangerous to him, if they may
be useful to man." In terms that Mary Shelley would have immediately
recognized, Bernard described a physiologist as "a man of science,
absorbed by the scientific idea which he pursues: he no longer hears the
cry of animals, he no longer sees the blood that flows, he sees only his
idea and perceives only organisms concealing problems which he in-
tends to solve."[31]

The new science of physiology first transformed the teaching of
medicine and then soon, often too soon, altered the practice of medicine. Before physiology, attending medical school was not very different
from apprenticing to a local physician, and in fact, state medical licens-
ing boards did not distinguish between the two forms of training. In
both settings, students learned the basic lessons of empirical medicine,
that is, what to do at the bedside. But once physiology became a core of
the curriculum, medical schools turned into centers for the teaching
of science. They took inspiration and guidance from Bernard's dictum: "Experimental physiology is the most scientific part of medicine
and . . . in studying it, young physicians will acquire scientific habits."[32]
Thus, when Johns Hopkins medical school designed a new curriculum
in the 1890s, it put physiology at the center of it, and recruited almost
all of its non-clinical faculty from its ranks. The Johns Hopkins model
spread quickly, and by 1920 physiology was the essential discipline in
university-based medical schools.

Physiology changed not only medical education but medical research
and practice. One of its earliest and most dramatic contributions was to
create the field of endocrinology. "The endocrines are fundamentally
basic to all principles of physiology," one investigator later noted. "In
fact, endocrinology is physiology."[33] The knowledge that glands produced and circulated vital substances through the bloodstream had been
suspected by some ancient and early modern researchers, but not until
Bernard was the idea confirmed. The closing decades of the nineteenth
century witnessed a flurry of research into these substances, and in 1903,
Ernest Starling, professor of physiology at London's University College, coined the term *hormone* (taken from the Greek *horman*, to arouse
or excite) to identify them. They were "chemical messengers . . . car-

ried from the organ where they are produced to the organ which they affect, by means of the bloodstream, and the continually recurring physiological needs of the organism must determine their repeated production and circulation through the body."[34]

The pace of research quickened over the 1910s and 1920s. In 1917, the inaugural issue of *Endocrinology* appeared, the first American journal devoted to hormones, and its editors boasted of the extraordinary amount of research under way. They estimated that between 1890 and 1917, some 15,000 articles on hormones had been published, casting new light on the causes and treatment of a variety of physical and mental diseases.[35] By the 1920s, hormones were occupying a place very much like genes today. As R. G. Hoskins, a leading Harvard researcher, declared, hormones have "a potency [which] is almost unbelievable. Their influence is pervasive in all that we do and are." Or as one of his colleagues put it: hormones "shape our skulls and they shape our souls. . . . They exercise a regulatory control over every aspect of the life of the individual from birth to death. . . . They make him and break him biologically and psychologically."[36]

Investigators diligently explored the biological impact of hormones, attempting to distinguish one substance from another and analyze its particular contribution. The task was complicated because the substances were very difficult to isolate. Nevertheless, knowledge gradually accumulated through imaginative, and highly invasive, animal research. Investigators pursued three principal strategies. First, they surgically removed a gland from an animal and then studied the aftereffects. Was the animal able to survive the loss? If it did survive, which physical capacities, if any, were affected or altered? This approach essentially turned the laboratory into a site for "the experimental production of pathologic conditions" in order to allow the investigator to study them.[37] Second, they attempted to isolate the substance that a gland produced— in this early phase, by the crude method of grinding up the gland that produced it and feeding or injecting the mixture into another animal to see what effects it might have. Third, researchers removed a gland from one animal and transplanted it to another animal, from the same species or across species (Dr. Moreau repeated), so as to learn more about its biological properties.

What knowledge came from these experiments? Physiologists discovered that hormones released by the pituitary gland, located at the base of the brain, affected body height. The composition of the chemical substances involved and the precise relationship between secretion and size eluded them, but they did understand that height depended not only on an individual's physical history (including his nutrition and health) and his inherited traits (how tall his parents were) but also on the hormonal production of the pituitary gland. Neurosurgeon Harvey Cushing at Johns Hopkins was responsible for some of the most important advances, demonstrating in 1909 through his work on puppies that the pituitary gland was essential to physical growth. Excise this gland from a puppy and it would fail to grow. By the same token, "diminished activity" of the pituitary gland in humans retarded physical growth and sexual development. Cushing also helped establish that "a pathologically increased activity" of the pituitary gland led to "overgrowth—giantism, when originating in youth, acromegaly when originating in adult life."[38] Moving from laboratory to operating theater, he removed one-third of the pituitary gland of a patient with acromegaly and reported that the procedure alleviated many of his symptoms.

Following on Cushing's work, Philip Smith, his Johns Hopkins colleague, conducted experiments on tadpoles and rats in order to map the structure of the pituitary gland and to demonstrate that excretions only from its anterior portion affected growth. Then, in a stunning demonstration of the hormone's power, Herbert Evans and Joseph Long at the University of California took anterior pituitary lobes that Armour & Company and Parke-Davis & Company had collected for them from cattle at slaughter (as we will see, a thin line distinguished meat-packing companies from drug companies), made them into a solution, and injected it into newborn rats. Within two months, the rats grew into giant rats, weighing 20 percent more than the standard breed.[39] Science was thus beginning to demonstrate its capacity to alter nature. As Cushing, only half jokingly, commented: "The Lewis Carroll of today would have Alice nibble from a pituitary mushroom . . . and presto! She is any height desired."[40]

Endocrinologists also discovered the special biological role of the thyroid gland, particularly how a deficiency in the hormone caused cretinism or the allied disease of myxedema, both of which are character-ized by physical and mental underdevelopment. The facial features of these patients were masklike, their skin was dry and thick, their mental capacities were diminished, their personalities phlegmatic, and their voices flat and monotonous.[41] "No type of human transformation," observed the great Johns Hopkins clinician William Osler, "is more dis-tressing to look at than an aggravated case of cretinism. . . . The stunted nature, the semi-bestial aspect, the blubber lips, retroussé nose sunken at the root, the wide-open mouth, the lolling tongue, the small eyes, half closed with swollen lids, the stolid expressionless face, the squat figure, the muddy dry skin, combine to make the picture of what has been well-termed the 'pariah of nature.' "[42] Imagine, then, the excitement that greeted the finding that an extract of the thyroid gland, initially taken from sheep and then, in 1927, synthesized in the laboratory, cured this wretched condition. Pariahs became normal human beings. George Murray, the English physician and pathologist, was the first to adminis-ter the hormone to a patient, helping to establish the value of "organ-otherapy," or what became endocrinology. Osler compared Murray's achievement to mythological fairies who performed magical cures by waving a wand or bestowing a kiss. Murray cured "unfortunate victims, doomed heretofore to live in hopeless imbecility, an unspeakable afflic-tion to their parents and to their relatives."[43]

The hormone that captured the most professional and public atten-tion was insulin. In 1921, two young Canadian physicians, Frederick Banting and Charles Best, demonstrated that a deficiency of insulin, a hormone secreted by the pancreas gland, caused diabetes. Heretofore, diabetics, in order to stay alive, had to severely restrict their food intake, tottering on the brink of starvation; because the balance between nutri-tional adequacy and a deadly excess of sugar could not be maintained for very long, diabetics typically died at a young age. Once Banting and Best uncovered the function of insulin and were able to collect the hor-mone from the pancreas glands of dogs, a fatal disease turned into a chronic condition. By taking their daily injections of insulin, patients returned to normal—it was Ezekiel's dream of dead bones coming to life. The success of the treatment was so spectacular that it won respect and admiration for the entire field of endocrinology. "If endocrinology had no other triumph to its credit than the discovery of insulin," biologists agreed, "that alone would be a sufficient reward for all the labors that have brought science to its present state."[44] Indeed, insulin opened wide the possibility of a whole series of miraculous hormonal substances. From the vantage point of the 1920s, bacteriology had already experienced its great leap forward—"infectious diseases are being conquered"—and the next great medical frontier apparently lay with endocrinology.[45]

The impact of these medical triumphs on social attitudes was twoedged. Advances in endocrinology encouraged and popularized a view of human nature that was essentially reductionist, maximizing the influence of biology and minimizing the impact of the environment on individual traits. Nature was elevated over nurture. You are your hormones, a proposition that was offered with the same confidence and conviction that would later be echoed in "you are your genes." But along with the reductionism of endocrinology came an intriguing twist: the study of hormones made human biology appear more flexible and open to manipulation, not immutable and permanently fixed. The initial discoveries held out the promise that endocrinology would empower medicine both to correct gross abnormalities, whether they were physical, mental, or behavioral, and then also to improve normal functioning. Hormones might be able to refashion the self.

II

Whatever fears the unbounded ambitions of science might generate, the therapeutic successes that endocrinology enjoyed, as well as its potential to improve physical and mental capacities, captured public imagination. Not that advances in biology were easily understood or quickly assimilated into an individual's view of himself and the world. In the 1870s and 1880s, for example, Americans had to absorb the counterintuitive if not downright preposterous idea that tiny organisms, visible to the eye only when stained and viewed under a microscope, caused such dangerous infectious diseases as tuberculosis, diphtheria, rabies, and scarlet fever. Then, in the 1920s and 1930s, they had to reckon with the fact that glands released secretions into the bloodstream that maintained health, warded off disease, and, even more astonishing, shaped the body and influenced the mind.

Information about endocrinology spread quickly to the public, but it still took time for the idea that ductless glands secreted vital substances into the bloodstream to be appreciated. The basic terms—*ductless glands, secretions*—were unfamiliar. Newspapers did not yet have medical journalists on staff, and drug companies advertised only in medical journals. The books that aimed to guide patients into the world of medicine had a very practical quality to them, filled with recommendations on what to do until the doctor came to treat an acute illness or injury. Chronic ailments, which constituted the substance of endocrinology, were generally ignored. Still, news of medical advances did circulate, albeit in odd ways.

One path was the human interest story with a medical twist. Americans had learned, for example, about Louis Pasteur's new vaccine to prevent rabies only when two New York children, bitten by a pack of wild dogs, were sent to Paris to receive the vaccine. The press covered the journey very closely, reporting on the children's arrival and their treatment by Pasteur, which, perforce, brought with it an explanation of his research and accomplishments. For endocrinology, the compelling human interest angle came from the circus freak, the performer in the sideshow who was considered to be an anomaly of nature. Writing in *The New Yorker* in 1934, Alva Johnson presented a three-part series on the freak shows, tracing their history from P. T. Barnum's circus, which exhibited Tom Thumb, to the more ordinary carnivals that featured the World's Ugliest Woman or the World's Biggest Man. Johnson meticulously described how entrepreneurs recruited their sideshow performers from places like Hamburg, Germany, and how midway barkers composed and delivered their "come and see it" spiels. But what intrigued him most were the so-called freaks themselves. He distinguished between the "authentic freaks," those born with their condition, and "synthetic freaks," like the tattooed woman, the glass eater, the fire eater, and the man who walked on coals, all of whom trained themselves for their circus performance.

The authentic freaks, Johnson explained, were not "monsters" or living representations of God's punitive judgment but individuals with a hormone deficiency or excess. Giving readers a crash course in endocrinology, he noted that "Modern medicine has shown that giants, midgets, fat ladies, and living skeletons are gland cases. They are born with abnormal thyroids, pituitaries, or other endocrine glands." These glands "produce powerful chemicals which affect the developing human organisms in many ways." Some of this new knowledge came from the freaks themselves, for they often "ended their careers on the dissecting table." Medicine now recognized that an excess of one chemical messenger could accelerate growth and produce giants, and a scarcity of it could retard growth and produce midgets. "Freaks," in other words, were patients with glandular irregularities who required medical care.[46]

The circus was also the vehicle that took endocrinology into elementary and high school hygiene and biology courses. Textbooks in the 1920s, and in more expanded versions in the 1930s, included a section on ductless glands, giving prominence to adrenaline (the hormone of flight) and insulin. But the center of the endocrine story went to the freaks, for who could resist using the bait of the giant and the midget to capture student attention?

"Have you ever strolled through a circus menagerie and seen the freaks?" asked the 1928 high school biology text *Health Essentials*, in its chapter "Startling Discoveries in Physical and Mental Development." "There is the fat woman, who is nineteen years of age and weighs three hundred and fifty pounds. There is the tiny Tom Thumb, who . . . is only thirty-two inches high. How the giant towers up more than eight feet tall!" To make certain that no young reader would miss the point, the text included a photograph of a giant beside a midget along with the caption: "The circus is in town. The greatest feature is the 'World's tallest and shortest man.' " Confident that students would be curious as to "Why are these people so different from others," the text dismissed the idea of abnormal growth as "a queer freak of nature" and introduced instead "the character of glandular excretions." An overactive pituitary gland produced the giant, an underactive one, the midget.[47]

Elementary school texts, such as *My Body and How It Works*, also invoked "some peculiar-looking creatures" to frame its discussion of hormones. "On the day when Jean and Jerry and John had been at the circus they had gone into the side shows." They came upon the fat lady ("the fattest thing you ever saw . . . as fat as a hippopotamus"), the giant (looking like a giraffe), and the "funny little dwarfs." Why had these people "not grown up normally the way most people do?" The answer came in a discussion of the "mysterious fluids" secreted by the endocrine glands and the consequences of their overproduction or underproduction.[48] Thus, students learned that medicine now had the ability to make the short taller and to prevent the already tall from becoming giants, establishing the idea that bodies could be shaped and reshaped.

No sooner was this idea glimpsed than a fascinating transition occurred in which cure moved into enhancement. Once the texts explained that endocrinology could redesign the body, it did not require a great leap of imagination to ask whether, if the short person could be made normal, could the normal person be made taller, and whether science someday might be able to alter not only our bodies but our personalities. The questions were so logical that one high school text concluded its section on hormones by raising them for class discussion: "Do you think it would be a splendid achievement if someday the scientist could modify our dispositions by simply injecting an extract which contained the desired personal quality?" A good question, not at all surprising when asked today, but startling to read in a 1920s high school text. Clearly, once the biological properties of hormones were recognized, the possibilities for enhancement became obvious.[49]

III

Novels in this period also explored enhancement through endocrinology, their tone set not by Mary Shelley's dark vision but by H. G. Wells's exuberance. His 1904 book *The Food of the Gods* anticipated Cushing's quip about Alice eating a pituitary mushroom and growing, presto, into a giant. Even as physiologists were only beginning to understand the biological significance of hormones, Wells was already addressing the social implications of the new knowledge. *Food of the Gods* incorporated shrewd speculations on the social consequences of a science that could turn children into giants. Although the substance that accomplished the task, boomfood, was not a hormone, Wells invoked the "ductless glands" to give authenticity to his account. The researchers themselves, consistent with Wells's overall view of scientists, were thoroughly conventional in appearance and demeanor, living in a "narrow world" and "monastic seclusion." The single remarkable thing about them was that they had "the unfaltering littleness" of scientific men.[50] Nevertheless, whatever they lacked in physical stature they made up for in talent, able to transform an ordinary child into a Hercules.

Unlike his earlier essay on individual plasticity, the novel was sensitive to the dangers that might follow on scientific accomplishment. Boomfood produced not only giant people but also, as its crumbs spread, giant wasps and giant rats (the 1923 achievement of Herbert Evans foretold). In fact, it generated so much havoc that Wells warned: "Think of all that a harmless-looking discovery in chemistry could lead to."[51] The humans who consumed boomfood did become giants, but they were at times more oafish than intelligent, becoming objects of scorn and violence, and ultimately, of an unsuccessful effort to send them—where else?—to Africa.

These qualifications notwithstanding, *Food of the Gods* made a powerful case for turning science loose as the surest route to biological and social progress. Efforts to constrain research were inappropriate and futile. Echoing the monster's thoughts in *Frankenstein*, one of boomfood's inventors declared: "We do set forces in operation—*new* forces. We mustn't control them and nobody else *can*. . . . The thing is out of our hands." Or, as Wells's omniscient narrator declared: "In spite of prejudice, in spite of law and regulation, in spite of all that obstinate conservatism that lies at the base of the formal order of mankind, The Food of the Gods, once it had been set going, pursued its subtle and invincible progress."[52]

Wells had some hopes that science would right class wrongs and loosen the hold of conservatives on politics. If the body of the lower classes changed, so might the body politic. Wells has political conservatives opposing boomfood on the grounds that while it made people bigger, it did not "make them happier" or "more respectful to properly constituted authority." Indeed, fearing that feeding "children of the lower classes . . . a food of this sort will be to destroy their sense of proportion utterly," they founded a national society—Temperance in Growth and a National Society for the Preservation of the Proper Proportion of Things.[53] Wells's parody of British conservatism reflected his profound unwillingness to allow a respect for the natural to dictate policy. Forget about the proper proportion of things so as to make way for science.

By the novel's end, Wells was rhapsodic over the progress that giantism would bring—oafishness was forgotten and all that remained was the vision and capacity of science to perfect humankind. As one of this new breed predicted: "A time will come when littleness will have passed altogether out of the world of man. When giants shall go freely about this earth—their earth—doing continually greater and more splendid things. But that—is to come. We are not even the first generation of that—we are the first experiments."[54]

TWO

The Female Principle

N O QUEST MORE ENGAGED the biological sciences over the
course of the twentieth century than fathoming the mysteries of
the female body. What caused the cycles of menstruation? How was
menstruation linked to reproductive capacity? At what point in the
monthly cycle were women most fertile and how might infertile women
be made fertile? What sparked the menopause and why was it often
accompanied by debilitating symptoms? These elemental problems
were rendered even more important by a deeply embedded cultural presumption that women—but not men—were ruled by their anatomy.
Their bodies endowed them with special psychological and moral traits
that prescribed their societal responsibilities. This maxim carried such
weight that even the new disciplines of physiology and endocrinology,
which finally answered the fundamental questions and transformed the
understanding of female biology, did not challenge it. Whether the
issue was fertility or menopause, ideas on what women should be were
enmeshed with what they were.[1]

What they were, as physiologists along with endocrinologists and
gynecologists explained, was rooted in the functioning of the ovaries,
not the uterus, as had previously been believed. By the 1920s, the
ovaries had come to represent the essential female organ because they
secreted the hormones responsible for the onset of menstruation, the
development of secondary sex characteristics, reproductive capacity,
and eventually, when the hormones were no longer secreted, for the
onset of menopause. "The ovary," as the new formulation posed it, "is

a most important hormonal organ in the female. Without it she is sexless, shapeless, sterile; her emotional and mental balance is remarkably disturbed; in sum, her identity as a woman is lost."[2]

Since the object under study was the female body, the findings of physiology had implications well beyond the laboratory. The earlier emphasis on the uterus had made femininity appear fixed and static, embedded in the organ that nurtured and protected the fetus from the moment of conception until birth. Anatomists diligently studied its physical structure (located deep in the pelvis, weighing 30 to 40 grams) and recognized some of its pathologies (cancer of the uterus, a tipped uterus). And they, and almost everyone else, identified the uterus as the part of the female anatomy that dictated women's destiny. As one medical text observed, "Women owe their manner of being to their organs of generation and especially to the uterus." Or, as an American physician asserted even more grandiosely: "It was as if the Almighty, in creating the female sex, had taken the uterus and built up a woman around it."[3]

By contrast, defining women by their hormones opened wide the possibilities for change. In this new paradigm, biology had the power to manipulate nature, to correct for its errors and also to improve on its design. To be sure, it was initially science and medicine that controlled the enterprise, instructing women on the choices that they now had. But eventually the hormonal concept of the female body served to empower women, allowing them to decide how or whether to make themselves over. As one sociologist trenchantly notes, the body turned into "a means for self-expression, for becoming who we would most like to be. In an era where the individual has become responsible for his or her own fate, the body is just one more feature in a person's 'identity project.' "[4]

I

Unlocking the secrets of the ovaries took time, energy, hubris, and an uncompromising readiness to ignore prevailing aesthetic or ideological sensibilities. Already in the seventeenth century, scientists had described the anatomical features of the ovary, but they did not grasp its biological functions. In 1672, Regnier de Graaf, a Dutch anatomist, observed round, fluid-filled chambers in the ovaries but what these chambers (later known as Graafian follicles) produced eluded him. A century and a half later, in 1827, Karl Ernst von Baer, a German professor of zoology, located inside these follicles small specks that were eggs. But neither he nor anyone else understood the relationship of follicle to egg, how the egg moved to the uterus, or what the biological relationship was between the ovaries and uterus.[5] Zoologists, and physicians as well, did associate the ovaries with sexual excitement and menstruation. Observing the seasonal mating habits of animals, they linked menstruation with coming into heat, oestrous as it was called, perhaps originating in a "nervous reflex" sent by the ovaries. They also believed that diseased ovaries stimulated excessive sexual urges, a presumption that led some surgeons, without a scintilla of evidence, to remove the ovaries of women suffering from nervousness, anxiety, or an unwillingness to perform household duties.[6] None of these preliminary observations or practices, however, detracted from the centrality of the uterus.

Only when physiology elevated the study of function over structure did the ovaries become essential to female biology. "It has been completely wrong to regard the uterus as the characteristic organ," observed Rudolf Virchow, one of the most influential German physiologists in 1856. "The womb, as part of the sexual canal . . . is merely an organ of secondary importance. Remove the ovary, and we shall have before us a masculine woman . . . [an] ugly half-form with the coarse and harsh form . . . the heavy bone formation, the moustache, the rough voice . . . the sour and egotistic mentality and the distorted outlook. . . . In short, all that we admire and respect in woman as womanly, is merely dependent on her ovaries."[7] Darwin's theories also contributed to the change. The meaning he gave to reproductive advantages in the survival of a species inspired greater scientific interest in the ovaries. Moreover, Darwin focused attention on the puzzling changes in secondary sex characteristics that occurred over a lifetime. Why did an aging female deer acquire horns? Why did an aging hen resemble a cock?

To find answers, physiologists together with zoologists dissected the ovaries of rodents, birds, and mammals to analyze, as one of them noted, the "extensive chemical changes that take place in these animals during the period of their reproductive activity."[8] With no deference to popular conventions of the "natural," they transplanted ovaries from one body site to another, from one species to another, and from mature animals to young ones, all in a relentless effort to discover the sources of sexual receptivity, menstruation, and secondary sexual characteristics.

Dissection and intra- and cross-species transplantation, although controversial, advanced knowledge. In 1896, Emil Knauer, an Austrian physiologist and gynecologist, grafted the ovaries of adult guinea pigs to the muscle of their abdominal wall and found that the ovaries continued to function; even when attached to a new site, they were able to prevent "castrate atrophy," as measured by the size of the uterus and vagina (which always shrank after ovaries were removed). Evidently, the transplanted ovaries continued to secrete a female-preserving product directly into the bloodstream. That same year, another Austrian, Josef Halban, grafted fragments of ovaries removed from adult guinea pigs into infant guinea pigs and observed that the infant uterus quickly grew to adult size. "We must assume," he wrote, "that a substance is produced by the ovary, which when taken into the blood is able to exert a specific influence upon the genital organs; and that the presence of this substance in the body is absolutely necessary for the maintenance— and, as my researches show, for the development—of the other genital organs and the mammary glands."[9] Although he could not isolate the substance itself, he was convinced that the ovaries were releasing a female "principle."[10]

Each discovery explained a little more about female sexual development and reproduction. In 1905, F. H. A. Marshall, a British physiologist, reported on the basis of his animal studies that the ovaries secreted not one but two substances, follicular (what would later be labeled estrogen) and luteal (progesterone). Both were essential to the process of menstruation, to initiating pregnancy, and to sustaining the fetus in its early stages. Marshall also demonstrated that administering ovarian secretions to a castrated animal would bring her into heat.

By 1910, the field had advanced to the point that Marshall needed 700 pages to complete a textbook on *The Physiology of Reproduction*. The existing literature, his preface observed, while replete with descriptions

of the digestive and nervous systems, had for too long ignored reproduction or consigned it "to a few final pages seldom free from error."
His text intended to fill the gap, analyzing, as befit a physiologist, life
processes, not anatomy. Marshall's chapters addressed such topics as:
"Changes in the Ovary"; "Changes that Occur in the Non-Pregnant
Uterus During the Oestrous Cycle"; and "The Changes in the Maternal
Organism During Pregnancy." All confirmed that the ovaries were the
source of female biological identity and capacities. To be sure, almost
all his data came from animal studies, but Marshall was certain that
his findings were highly relevant to "medical men engaged in gynaeco-
logical practice. . . . Generative physiology forms the basis of gynaeco-
logical science, and must ever bear a close relation to the study of
animal breeding."[11] The hen's egg had much to teach us about human
reproduction.

Any lingering doubts about the central role of ovarian secretions
were rebutted by the startling and imaginative research carried out by
Frank Lillie, a zoologist at the University of Chicago. Lillie studied the
embryological development of the bovine freemartin, an odd creature
in nature, the sterile female twin to a male calf. He discovered that the
freemartin was conceived and implanted as a separate egg but then, at
an early stage of fetal development, its chorion, or fetal membrane,
fused in the embryo with the male's and became permeable. The blood
of the two embryos mixed so that the female freemartin received the
blood-borne secretions produced by the male freemartin. Because the
female embryonic development was "modified by the sex hormones of
the male twin," she was born sterile, in effect, half-male, half-female.
Thus, Lillie concluded: "The course of embryonic sex-differentiation is
largely determined by sex-hormones circulating in the blood."[12]

II

From its inception, physiology was an empirical science, with investigators ready to see scientific knowledge gained through laboratory experiments brought to the bedside. Clinicians, in turn, were ready to make
patients available to physiologists.[13] "The wards," declared William

Osler after visiting the laboratories of prominent German physiologists, "are clinical laboratories utilized for the scientific study and treatment of disease, and the assistants, under the direction of the professor, carry on investigations and aid in instruction."[14] Despite a strong consensus on applying laboratory findings to patient care, several very controversial questions emerged: What level of evidence was necessary before laboratory findings were translated into bedside practice? On whose authority and by what standards of proof should experimental results become the basis for therapeutic interventions? With some notable exceptions, physiologists gave conservative answers; all too often, clinicians did not. They frequently dispensed hastily developed and often rudimentary compounds to their patients, paying little attention to the purity of the substances or the adverse effects they might have. Although random clinical trials had not yet become the gold standard and governmental bodies were not regulating drug safety and efficacy, still the members of a profession who had sworn to do no harm were often reckless in dispensing substances of doubtful benefits and unknown risks.

This recklessness dogged almost all the applications of would-be hormonal therapies. The story begins with Charles Édouard Brown-Séquard, a commanding figure in experimental physiology who had succeeded Claude Bernard as the chair of medicine of the Collège de France. In 1889, at the age of seventy-two, he startled his colleagues with two announcements. The first and most dramatic, which we will return to later, was that he had injected himself with a mixture of testicular extracts and apparently experienced a remarkable invigoration. He was no longer tired, weak, sleepless, or constipated. His second claim, presented with a little more reserve since it was not based on self-experimentation, was that "the ovaries of animals must produce an extract which has the same effect on women as the testicular extracts have on men."[15] A Paris midwife, Brown-Séquard reported, had found her health and energy restored after swallowing a liquid made from guinea pigs' ovaries. So too, a Marseilles surgeon claimed he cured a patient suffering from hysterical seizures by administering ovarian extracts after the removal of her ovaries. An American physician, Augusta Brown, whom Brown-Séquard described as a woman of great courage, had injected forty-six elderly women with an extract made from the ovaries of rabbits. It not only diminished symptoms of hysteria but also reduced the "debility due to age."[16] Brown-Séquard was enthusiastic about these findings, although he did admit to some uncertainty as to whether ovarian extracts were as potent or "specific" as testicular extracts.

Given the fact that physiologists were only beginning to understand what hormones were, the diversity and range of reports that Brown-Séquard cited was astonishing. Already by 1889, clinicians, surgeons, and midwives, both in the United States and Europe, were administering different forms of ovarian extracts to treat everything from hysteria to aging. Well in advance of isolating the substance (which would not occur until 1923), or understanding how it worked or exploring the risks that its use might pose, a group of physicians—soon aided by drug companies—was removing, crushing, and liquefying animal ovaries and then administering the mixture to women patients.[17]

Although general physicians dispensed these extracts, it was the enthusiasm of gynecologists that established their clinical primacy. Gynecologists helped legitimize ovarian therapies and, in turn, ovarian therapies helped legitimize gynecology. By the time of Brown-Séquard's announcement, the specialty was well established in Europe and the United States, attracting middle-class patients and determined to attract still more. To this end, gynecologists deliberately associated themselves with laboratory science; once physiology vouched for the power of the "female principle" and made the ovaries the critical female organ, they moved rapidly to translate this finding into clinical practice. They also welcomed a novel therapeutic agent. They were convinced by both medical theory and clinical experience that women, far more than men, were prone to illness, that good health for them was never more than a temporary, really illusory, phenomenon. For women, as one gynecologist explained, "there is no proper boundary between physiology and pathology . . . pathology is but a chapter in the history of physiology."[18] Viewing their women patients as essentially pathological, they were prepared to dispense new therapies, proven or not, risky or not.

So too, ovarian extracts provided gynecologists with an alternative to the surgical removal of the ovaries, which they called *ovariotomy*. They had been using this highly invasive and dangerous procedure not only to treat ovarian tumors, fibroids, and cysts but also to relieve pelvic pain, hysteria, and even epilepsy. The surgery, however, was becoming increasingly controversial, with a growing number of critics in the 1880s denouncing practitioners as "unscrupulous or illogically enthusiastic experimenters." "We might at least require some evidence of the ovaries being diseased," complained one prominent opponent, "before consenting to their extirpation in the hope of curing any of those vague disorders to which women are so subject."[19] As the debate over the procedure intensified, the incentive for gynecologists to adopt glandular therapies increased.

For all these reasons, ovarian extracts became a popular medical remedy and by the mid-1890s a spate of articles in gynecological journals described their purported benefits. German gynecologists at the Landau clinic in Berlin were especially enthusiastic. A typical case report from the clinic described how ovarian extracts relieved episodes of hot flashes, sweats, and giddiness in a twenty-three-year-old woman whose ovaries had been surgically removed. As soon as treatment began, the episodes diminished; when the therapy was stopped or a placebo substituted, the symptoms reappeared. "Ovarian extract," concluded Leopold Landau, the clinic's founder, in 1896, "is a remedy that enables us to diminish the severe troubles of the natural or artificial menopause, and usually to cure them, without any unpleasant results."[20] French gynecologists were no less confident of its efficacy. An 1898 text, *The Ovarian Secretion*, declared that with the administration of a combination of desiccated ovary glands and fresh sow glands the "troubles of menopause . . . disappeared or were much ameliorated . . . without other medication."[21] The first English-language monograph on glandular therapy, published in 1905 by a British gynecologist, affirmed that ovarian substances effectively relieved menopausal women whose ovaries had "atrophied."[22] Enthusiasm was not dampened by the diversity of compounds that were on the market or the difficulty of knowing precisely what they contained. In short order, the extracts would be used to treat amenorrhea, dysmenorrhea, painful sexual intercourse, and sexual frigidity.[23]

The drug companies did their best to build a substantial market for hormones, however adulterated or ineffective they were. In this pre-1920 period, the bedrock of drug companies' sales and profits were patent medicines—a misnomer if ever there was one because the remedies were purposely not patented. Given the inferior and haphazard quality of the preparations, the companies preferred to keep the formulas secret rather than list ingredients and receive, in return, exclusive rights to marketing them for a number of years. Some companies used ground-up ovaries from hogs, others used cows, and still others mixed the two. Some compounds were made from the entire ovary of an animal, while others contained fragments turned into a juice.[24] Some preparations were made after removing the corpus luteum from the ovary; others ground it into the mixture. Some companies claimed to use only ovaries extracted in an early stage of the oestrous cycle, because they were more potent. Competitors insisted that the most effective extracts were those taken from animals in heat, or from pregnant animals.[25] Needless to say, no data existed to back up any of these claims or to know whether the preparation actually contained what the label suggested. Only a few physicians, however, made these points in print and drug companies did their best to ignore them.

Some physicians experimented with alternative therapies that were even more invasive and no less unproven. A number of surgeons attempted to transplant ovaries from healthy women to women who were unable to menstruate in an effort to restore menstrual functions. But the techniques for successful transplants were well beyond them even if the purchase of healthy ovaries was not. Another approach, followed by Eugen Steinach and his disciple Harry Benjamin (both of whom, as we will see, were more interested in the male hormone), applied X rays to the ovaries in an effort to stimulate them to greater productivity.[26] But (thankfully) neither of these two procedures spread far within the profession.

To be sure, a number of gynecologists urged a modicum of caution. They advised colleagues to choose carefully among brands or to mix their own extracts following published instructions. But choosing carefully was not really an option, since no one knew whether one brand was really more reliable or efficacious than another or what any of them actually contained. Moreover, the medical belief was that the extract, whatever its composition, had to be taken for weeks or months to be effective and improvements would not persist unless treatment continued. But again, ignorance and confusion reigned for no one was certain when to start, or when to stop, treatment.[27] Even negative results had little impact because gynecologists confronting patients who did not improve blamed failure on the patient (as in, the patient failed the therapy) or switched to another compound or regimen.[28]

The popularity of ovarian extracts was so widespread that even highly trained and sophisticated investigators, wary of using products of unknown quality and efficacy, dispensed them. In 1922, Emil Novak, a member of the Gynecology Department at Johns Hopkins, complained that "in judging the results of ovarian therapy . . . most of us have unconsciously succumbed to the familiar 'post hoc, propter hoc,' error into which it is so easy to slip."[29] If the treatment worked, then the patient must have been suffering from an ovarian deficiency; if it did not, then she must have had a different problem. Novak fully appreciated just how feeble the evidence was for efficacy. Yes, the removal of the ovaries did halt menstrual function and could bring painful symptoms. "It does not follow, however, that the feeding or the injection of ovarian extracts can replace the normal ovarian secretion, that it can restore the menstrual function and cause a cessation of the menopausal symptoms."[30]

Nevertheless, despite his frank criticism of faulty logic and drug unreliability, Novak, too, dispensed ovarian extracts regularly, resorting to a tortured defense of hormonal therapy. "I believe it to be based on rational principles," he declared, and therefore, "it is reasonable to hope that the biological chemist will sooner or later succeed in giving us ovarian extracts which will really approximate in their effects the action of the ovarian secretion intra vitam."[31] Because it should work and someday will work, it now does work—illogical on the face of it and yet, then and still now, regularly repeated. Whether for reasons of hope, desperation, or income, ovarian therapy was too alluring for physicians to resist. Better to give something even if it was, therapeutically speaking, nothing.

III

As clinicians were complacently dispensing these dubious compounds, physiologists, together with their chemist colleagues, persisted in their attempts to identify, purify, and synthesize the active ingredients in ovarian secretions. Turning the female principle into chemical form was far more daunting than they had anticipated.[32] Generalizing from the experience of organic chemists in Europe, they thought the task would be relatively easy. After all, German chemists had successfully identified and isolated an active ingredient in the bark of willow trees, salicylic acid, and produced a substance that was even more effective than oil of wintergreen in relieving arthritic aches and pains. The chemists had then gone on to synthesize a particularly potent version of the compound—acetyl-salicylic acid—thereby creating what became known as aspirin.[33] But deriving a compound from a tree and extracting it from an animal were markedly different assignments.

The first problem that investigators confronted was how to gain access to the quantities of raw materials they needed. It took ingenuity, diligence, and more—a hardheaded and unromantic readiness to pursue nature to its hiding places. There was no room for the philosophically or physically squeamish. The easy part of the assignment was to persuade universities with their newfound enthusiasm for research to build relatively large facilities for experimental biology. Within them, investigators had the space, both physical and moral, to conduct their experiments on guinea pigs and colonies of mice and rats.[34] As important as these species were for some experiments, they were not useful in the effort to collect ovarian secretions. Since the amount of the substance stored in the ovaries of any one animal was small, investigators required an exceptionally large number of ovaries removed from very sizable animals. Even so, it took literally tons of sow glands or cow glands to isolate a milligram or so of ovarian extracts.

To obtain their raw materials, researchers often contacted local slaughterhouses and packing plants, persuading their managers to save animal organs and glands. When the University of Chicago's Frank Lillie was researching the physiology of the freemartin, he required cow uteri containing twins. He arranged with the superintendent at Swift & Company at the Chicago stockyards to have his butchers notify him when they came upon such a uterus. The superintendent then called Lillie, who dispatched his assistant to collect it.[35]

Lillie's needs were simple compared to those of other investigators. He merely wanted to examine a single organ at a relatively easily recognized moment; no one had difficulty identifying a pregnant cow. For other researchers, however, who were attempting to understand the physiology of the menstrual cycle and the periodic flow of ovarian extracts, the problem of collection was far more complex. Hormonal secretions waxed and waned, and none of the changes were visible to the naked eye. In order to achieve successful results, therefore, these researchers were compelled to gather large quantities of material and carefully excise and preserve the ovarian tissue.

The more precise the questions that researchers proposed, the more important it was for them to be as proximate as possible to the abattoir. When Franklin Mall, chairman of the anatomy department at Johns Hopkins University, set out to study the embryonic development of pigs, he located his laboratory a block away from a Baltimore slaughterhouse.[36] Closeness to the research materials more than compensated for the sight and sound of herds arriving at the plant to be slaughtered and the stench of dead carcasses. So too, George Corner conducted his graduate school research into the physiological changes in the ovary over the menstrual cycle by persuading the owners of a San Francisco abattoir to let him use one of their shacks as a "field laboratory." Corner tagged the sows that were in heat and, immediately after they were slaughtered, sorted through their entrails to obtain the uterus and the ovarian tissue. He actually supervised the slaughter of the animals so as to ensure getting what he needed.[37] When Corner became a faculty member, he persuaded Johns Hopkins to pay for the purchase and special slaughter of twenty-two sows and instructed the butchers to slow down "the pace of the killing floor routine so that I could collect the ovaries and uteri systematically."[38]

Since the largest animals had the largest ovaries, Alan Parkes, a

British physiologist thought he had the game beaten when, with the help of the British Museum, he got access to the ovaries of southern blue whales—animals that can weigh up to 70 tons. Parkes thought his days of visiting the slaughterhouse were over, only to discover that the seamen had not carefully followed his instructions for preserving the tissues. By the time a barrelful reached his laboratory, it contained "only a mass of stinking material which we disposed of with some difficulty via the Institute incinerator amid complaints from people living near."[39]

Physical proximity along with the tools of chemistry eventually brought about the isolation and purification of the female principle. In a chance meeting between Edgar Allen, a physiologist at Washington University's School of Medicine at St. Louis, and Edward Doisy, his biochemist colleague, Doisy described his attempts to purify the insulin produced by the pancreas, and Allen shared his ideas on the ovaries as an organ of internal secretion.[40] They decided to conduct a joint investigation to isolate the ovarian secretion, and regularly drove to the local meatpacking plant, gathered as many ovaries from sows as they could fit in Doisy's Model T Ford, and brought them back to the laboratory. Since the slaughtered sow could be at any point in her three-week oestrous cycle, either before or after the follicle had matured, they could not be certain that they had collected the material they needed.[41] Still, using an alcohol solvent, they were able to extract some fluid (about 100 cc, or one hundredth of a gram) from each pound of ovaries; eventually 4 tons of sows' ovaries yielded them a few milligrams of the ovarian fluid. Injecting this fluid into spayed mice, they discovered that it caused changes in the vaginal tissue consistent with the mouse going into heat and that shortly thereafter the mouse would display the behavior typical of being in heat.[42] Once confident that they had isolated the active substance, Doisy, drawing on his experience in purifying insulin, extracted the proteins and lipids from the follicular fluid and through this purification process isolated a tiny amount of the active substance, that is, estrogen.[43]

Other investigators searched for accessible and plentiful sources of estrogen in different bodily fluids. They discovered that a sow's urine also contained estrogen, and contracted with farmers who were delighted to sell liquid waste for almost the same price as milk. In 1927,

Selmar Aschheim and Bernhard Zondek, two German investigators, discovered even larger amounts of estrogen excreted in the urine of pregnant women. This finding was particularly important because estrogen
derived from urine had neither proteins nor lipids and hence did
not require the complex extraction procedures that Doisy and Allen
had perfected.[44] "A watery source of the hormone," as one researcher
explained, "is very much easier to work with than follicle fluid or
chopped-up placenta."[45] Requiring less time and fewer materials, the
process facilitated the isolation of much larger quantities of uncontam-
inated estrogen. Doisy himself quickly took advantage of this breakthrough. He persuaded a nurse in the university's obstetrics clinic to
give each patient a gallon bottle with instructions to return it filled with
urine. When he needed even greater supplies, he had his graduate students deliver empty bottles to the women's homes and bring the filled
ones back to the laboratory. His persistence was rewarded. In 1929,
Doisy first, and then others in quick succession, produced crystalline
preparations of estrogen that were plentiful, pure, and potent.[46]

Ironically, the intense search for the female principle ultimately
blurred the sharp divide between male and female. In 1929, Bernhard
Zondek, after finding estrogen in the urine of pregnant mares, discovered, to his and others' astonishment, estrogen in even greater quantities
in the urine of stallions.[47] How could the fabled representative of masculinity be a rich repository of the female principle? The finding (which
reflected just how close in chemical composition the two hormones
were) was so counterintuitive that as the physiologist Herbert Evans
later remarked, an era of clarity—when men reputedly produced male
hormones and women, female hormones—gave way to an "epoch of
confusion."[48] Although the respective male and female hormones determined reproductive capacity and secondary sexual characteristics, sexual differences did not indicate an absolute hormonal division but a
different balance in the amount of each present. Male/female was a
matter of hormonal proportion.

The overriding lesson that science and medicine drew from these
findings, already suggested by Bernard, Loeb, Haldane, and Wells,
was that stripping nature of her secrets opened up vast possibilities
for biological manipulation. As distinctions once thought to be rigid

and unbending weakened, the feasibility of strengthening, restoring, or revising physiological conditions seemed within reach. Science and medicine were acquiring the knowledge not only to cure but to design and perfect.

IV

No sooner were new estrogen products available than endocrinologists and gynecologists extolled their use and dispensed them freely to their patients. A trial-and-error approach prevailed, not a cautious and careful accumulation of data on risks and benefits. "We can only learn by experience whether female sex hormone therapy will be of any value," insisted the noted Dutch physician and pharmacologist Ernst Laqueur. "Theoretically one cannot make the slightest prediction on whether it will have any useful effects. In the end this will have to appear in practice."[49] Emil Novak, who had complained in the 1920s that the quality of hormonal preparations was so low that "the results in all forms of ovarian therapy are unfortunately still very inconsistent," was delighted a decade later to reverse his opinion.[50] "We now have available many commercial preparations of unquestioned potency and reliability."[51]

The first powerful agent produced in pure and copious quantities was estradiol. Because of its chemical composition, however, it could only be given to patients by intramuscular injection. Women would have to pay their doctors for twice-weekly office visits that when combined with the cost of the drug made the yearly charge, in today's dollars, nearly $10,000. By the early 1940s, drug companies were producing more effective and less expensive versions of estrogen that could be taken orally.[52] They included diethylstilbestrol (DES), a synthetic estrogen developed by the English chemist Edward Dodds, and Premarin, an estrogen tablet made from the urine of pregnant mares.[53] These two products became the treatment of choice, making estrogen available to many more women.[54]

Estrogen, as Emil Novak observed, now assumed "greater importance in gynecological practice than in any other branch of medicine."[55] In fact, endocrinology and gynecology practically merged, encouraging gynecologists to explore almost every aspect of their patients' lives. "The mental, physical and sexual endowments of the healthy woman," remarked Edwin Hamblen, professor of gynecology at Duke and head of its sex-endocrine clinic, "are expressions of an endocrine symphony."[56] As his metaphor suggested, the gynecologist was the conductor who exerted his authority over his women players.

Thus, gynecologists began paying attention to a variety of anomalies that might be linked to ovarian malfunctioning. These included precocious puberty (the development of female secondary sex characteristics in young children), genital infantilism (the absence of secondary sexual characteristics in teenagers), forms of virilism (the bearded lady), and women of exceptional obesity (the fat lady).[57] Estrogen also helped to create the field of adolescent medicine, which brought gynecologists a patient they had rarely seen before, the young, unmarried girl. Adolescence, as one practitioner declared, ushered in "important endocrine metabolic, physical, and psychic adjustments."[58] Were there any deviations from normal development, gynecologists should intervene immediately; a symptom such as delayed onset of menstruation might signal an underlying pathology that could later prevent pregnancy. What precisely constituted delayed onset? No one could be certain. Menarche ordinarily began between the ages of twelve and fifteen, but the medical literature, drawing on anthropological findings, noted significant variation, reflecting differences in family income, race, ethnicity, diet, hygiene, and climate.[59] Nevertheless, it was preferable to move quickly rather than miss an opportunity. "A late menarche," warned two eminent gynecologists, "is the signpost of an adolescent endocrine disturbance . . . [and] mal-development of the lower genital tract."[60] Hormonal treatment had to begin "promptly and efficiently not later than at the age of 16 . . . if future impediments to genital function and fertility are to be avoided."[61]

No sooner did menstruation come under medical scrutiny than disease categories proliferated. For example, the traditional category of amenorrhea was divided and then subdivided into infantilism (underdevelopment of sexual organs), Fröhlich's syndrome (infantilism due to a pituitary dysfunction), primary amenorrhea (delayed or imperfect development of secondary sex characteristics), secondary amenorrhea (premature menopause), polymenorrhea (short menstrual cycle), hypomenorrhea (too little menstrual flow), oligomenorrhea (delayed menstruation due to pituitary dysfunction), and endocrinopathic amenorrhea and sterility (due to a dysfunction of the ovary, the thyroid, or the anterior pituitary gland). All of these categories might respond well to the administration of estrogen.[62]

Gynecologists now charted the regularity of patients' menstrual cycles, although no clear-cut definitions demarked normal from abnormal periods. "The interval between the onset of successive menstruation varies considerably even in the same individual," declared one major gynecological text. "There is no unanimity of opinion as to the most common average length of the menstrual cycle."[63] As another researcher succinctly observed: "The only regularity about the menses is their irregularity."[64] But gynecologists worried more about doing too little than about doing too much. "There was a time," Novak recalled, "when menstruation was looked upon as a pelvic phenomenon, involving the uterus and the ovaries alone. . . . We now conceive of the reproductive cycle as a constitutional one, involving organs and tissues far removed from the pelvis. . . . Menstruation is only one manifestation of this far-flung cycle."[65] All of which confirmed the need to do something.

Gynecologists were eager to treat the physical and psychological accompaniments of the menstrual cycle. In 1931, Robert Frank, reputed by colleagues to be a brilliant investigator, identified a syndrome that he labeled "premenstrual tension," and, thereby, transformed both gynecological practice and popular attitudes.[66] Following a group of women of childbearing age, Frank and his colleagues observed that the women regularly experienced "pathologic manifestations of some of the normal menstrual complaints" seven to ten days before menstruating; even more oddly, the complaints stopped immediately once menstruation began.[67] The symptoms included cyclic "alteration of personality . . . physical unrest . . . constant irritability . . . unreasonable emotional outbursts."[68] The most common symptom was a "feeling of indescribable tension."[69] Because the syndrome varied with the menstrual cycle, they hypothesized that the underlying cause had to be hormonal: women with premenstrual tension "do not excrete estrogen in a normal manner,"[70] and recommended estrogen to palliate the nervous system and restore equanimity, in the process bringing gynecology into the field of mental health. It was a short step, then, to using estrogen for improving women's sex lives. Although gynecologists conceded that many sexual difficulties reflected a loss of "libido" and, therefore, were the psychiatrist's turf, they maintained that symptoms of "frigidity" or "dyspareunia" (painful coitus) often had an endocrine cause and belonged under their own purview.[71]

Gynecologists had long treated women experiencing difficulties in conception; but once endocrinology entered their practice, they were even more ready to intervene. They thought that fertility and sterility lay along a spectrum. "Among nonsterile individuals," contended Harvard's R. G. Hoskins, "all degrees of fertility [appear] from the lowest to the highest. Not only does the grade of fertility vary from one individual to the next, but it may change from time to time in the same person, depending upon a variety of circumstances."[72] These variations suggested that infertility might be responsive to hormonal treatment. Gynecologists also believed that they had the ability to prevent the onset of fertility problems: "The ideal time to treat endocrine sterility," Hamblen noted, "is before it develops," so physicians should be prepared to administer hormonal treatments to their young patients.[73]

Were there data to confirm all these propositions? Assuredly not. Over these years, the Council on Pharmacy and Chemistry, established by the American Medical Association in 1905, reviewed drug products and evaluated their contents and company claims. (The AMA has always been a complicated and conflicted organization, sometimes working for the public good, as witness the Council on Pharmacy and Chemistry, sometimes working against it, as exemplified by its fierce resistance to the enactment of Medicare.) The council, made up primarily of investigators in pharmacology, not clinicians, was especially attentive to the new estrogenic compounds and repeatedly emphasized how little was known about their purported benefits or potential risks. "Great caution is necessary in the use of these preparations," it declared in 1933, "and greater caution in making deductions from it. The indiscriminate use is likely to do more harm than good. . . . The clinical use has kept far ahead of the laboratory data. . . . Most of the basic facts should first be worked out in the laboratory before they are tried in the clinic."[74] When DES appeared, the council was disturbed by the large percentage of patients who, in the first trials, had developed toxic reactions. Clearly more laboratory studies were needed, and until they were completed, "it should not be recognized for general use."[75] So too, when progesterone therapy emerged in the early 1940s, the council found it was of "limited value." The initial reports were "overenthusiastic" for so "experimental" an intervention.[76]

The council, however, had no regulatory authority and no ability to fund clinical research and collect evidence. It was also aware that its findings were regularly ignored by practitioners. As one of its consultants observed, the frequency of patient complaints and the absence of other remedies make it "easy to understand why the harassed clinician will continue to employ endocrine therapy in spite of its questionable value for many indications."[77] Moreover, clinicians had little interest in data on outcomes. "Therapy for endocrine disturbances," Robert Frank maintained, "can be looked at from one of two points of view: from that of one who insists on having definite proof in his therapy . . . or from that of one who tries out drugs and sees what the effect is."[78] Frank and many of his gynecological colleagues were in the second school, ready to tinker and see what happened.

Gynecologists responded to menopause in this same fashion. Some women at the cessation of menstruation complained of hot flashes, sweats, irritability, and moodiness. Just how many suffered these symptoms was unknown, since the evidence was essentially anecdotal. Emil Novak, who enjoyed a very large gynecological practice, estimated that three-quarters of his patients did not experience these complaints. To be sure, "a minority of menopausal women need medical treatment,"[79] and he gave them the new estrogen products. He was pleased with the outcome: "While formerly there was much skepticism as to the value of ovarian therapy for the relief of menopausal symptoms . . . there is no longer any doubt as to the results of present-day estrogenic therapy." With estrogen, "the typical vasomotor symptoms of the menopause can thereby almost always be kept under reasonable control, and the lot of the menopausal patient made a much happier one."[80] George Corner, as expert and cautious as anyone in the field, concurred. "When the characteristic symptoms of a stormy menopause, such as hot flashes of the skin and general nervousness, become almost intolerable, large doses of estrogenic hormones often give genuine relief."[81]

As gynecology expanded its domain, the category of disease became more elusive. It was one thing to correct a marked hormonal deficiency so as to correct genital infantilism. But what did "irritability" represent and what purpose was being met by giving women estrogen to combat it? Advances in endocrinology were making unhappiness a complaint that patients could properly bring to their physicians (not only psychiatrists) and receive interventions (that owed nothing to psychiatry). When Novak maintained that estrogen administered to menopausal patients made them "happier," he was charting a new direction for medicine, putting it on the road to enhancement.

Another striking example of this trend at work was gynecologists' response to women with primary amenorrhea, patients whose failure to menstruate was the result of ovarian malfunctioning. There was little that gynecologists could do to correct the ovarian defect and stimulate ovulation even with potent estrogenic substances. The drugs could produce uterine bleeding and produce minor cosmetic changes (a slight increase in breast size), but they had "no stimulating effect upon the ovary."[82] "No amount of estrogenic substance," Novak reported, "can produce in the endometrium the sequence of histologic changes characterizing the menstrual cycle." In fact, patients' prolonged use of estrogen actually inhibited ovarian activity. Hence, he concluded, "the treatment of amenorrhea by estrogenic substances is notoriously unsatisfactory."[83]

Nevertheless, gynecologists prescribed estrogen in these very cases, stimulating a bleeding that was, as Novak described it, "purely substitutional" and had no relationship to genuine ovulation.[84] "If these remarks appear unduly pessimistic," he declared, "I can say only that I know of no gynecologist of standing who is not unenthusiastic about the treatment of amenorrhea by estrogenic substances." But then Novak backtracked. "On the other hand, I know of few who do not often resort to [estrogen] for amenorrhea because they know of no treatment that offers any more prospect of success."[85] He himself belonged to this "treat anyway" camp. Although estrogen was "of little value in the treatment of endocrinopathic amenorrhea, . . . It is often resorted to because of the lack of any other treatment which is any more rational or any more effective."[86] In effect, gynecologists recognized that "treatment accomplishes nothing," that induced flow was unrelated to ovarian function. Why then did they administer estrogen? Their answer was to induce a flow that had a "favorable psychological effect on the patient." Women would be happier because they mistakenly believed that they were menstruating.

Take a case reported by Columbia endocrinologist, Raphael Kurzrok. When a very short and stocky sixteen-year-old girl who was sexually underdeveloped and had never menstruated came to him, he administered large doses of estrogen. After nine months, when she had experienced seven cycles of vaginal bleeding, Kurzrok upped the dosage and, although it produced only a few cosmetic changes (the patient's breasts came to "resemble those of a young adolescent"), he persisted. "The results obtained in this case (and in other similar cases) are satisfactory in that the patient feels like any other normal woman, in that she has breasts and 'menstruates.' To us the treatment leaves much to be desired, first because the effects obtained are temporary, and second, the treatment is solely replacement therapy, for there has been no permanent stimulation of the gonads." Nevertheless, "in spite of these shortcomings, the treatment is worthwhile from every viewpoint." The patient felt more and looked more like a woman.[87]

Physicians continued to prescribe estrogen for amenorrhea even though they understood that over time, the dose of estrogen had to be greatly increased to produce this cyclical bleeding, and that eventually even megadoses produced only infrequent and scanty periods. So even when a physician continued to administer estrogen and the patient was desperate enough to keep taking it, the pseudo-menstruation and a womanly appearance could not be maintained. Over time, the patient inevitably reverted to her earlier state.

This endocrine masquerade served physicians well in several ways. First, they were well paid for the visits and the drugs. Second, they believed that intervening in this fashion put them in the vanguard of medical progress. "We know that the results that we have obtained are reversible," Kurzrok explained. "The patients revert to their original status of sexual development upon cessation of treatment. This may be unsatisfactory to some clinicians, but we feel that a beginning must be made. . . . The next few years will see the production of more potent gonadotropic hormones . . . with prolonged and constant action."[88] Third, physicians took comfort in the idea that patients felt better even if they were not better. At least for a brief time, estrogen reduced the sadness that came with the absence of menstruation.

One last consideration must be added to the story. Patients were made happy but at what level of risk? The most prominent one, well appreciated in the late 1930s and early 1940s, was the possibility that estrogen use caused cancer. The data was not conclusive but at least some researchers had a high index of suspicion, which they expressed in the major journals. One *Journal of the American Medical Association (JAMA)* editorial in 1940 opened by noting that "malignant growths of the mammary gland and uterus have been induced in rodents with estrogens by numerous investigators." It then cited one case report of a patient with a family history of breast cancer who developed "a mammary cancer following prolonged use of estrogens," and a study that associated estrogen use with cancer of the cervix. The editorial even warned that decreasing the amount of estrogen given might not obviate the risk "since it is quite possible that the same harm may be obtained through the use of small doses of estrogen if they are maintained over a long period."[89] The lack of conclusive data kept the editorial from suggesting that physicians never give any women estrogen but it urged physicians to be particularly cautious with patients who had family histories of cancer.

In this same vein, the investigators who evaluated DES for the council reported (again in *JAMA*) that this compound alleviated symptoms of menopause. But they cautioned that estrogens were not "innocuous substances but act on various organs and functions of the body." In animals, they had cancer-causing effects that "have not been as yet clearly demonstrated in the human being." Still, these findings "should raise a reasonable doubt in the mind of the clinician that these substances may be given with impunity." Physicians certainly should not give long courses of DES to patients with a family history of cancer. In all cases, they should "watch for latent posttherapeutic results."[90]

Caution and watchfulness, however, did not mark clinical behavior. Emil Novak, for example, in 1933 acknowledged that "estrogenic substances administered in large doses and for prolonged periods in certain experimental animals, have been shown to produce cancer . . . and has led to the fear that excessive estrogenic therapy may be attended with a risk inciting the development of cancer in the treated patient." But the evidence, he insisted, was inconclusive, and depriving women of estrogen because of the "slight theoretical possibility" of cancer was "carrying conservatism and caution to an extreme."[91] A decade later, he had not changed his position: "There is practically no evidence as to this supposed danger from the standpoint of the human. It is true that one or two cases have been published in which estrogen therapy has been followed by the later development of cancer, but their significance appears very unconvincing. The doses of estrogen required to produce cancer in animals have been huge." Novak conceded that despite "no worthwhile evidence," physicians would be well advised in patients with family histories of the disease or with "any so-called precancerous condition . . . to avoid or to limit estrogenic therapy as far as possible."[92] But otherwise, estrogen could be dispensed without fear of risk.

Other gynecologists took the same position. "The available evidence relating to the role of estrogen in the etiology of human cancer," insisted Philadelphia gynecologist Jacob Hoffman, "is limited and inconclusive. . . . The ability of the estrogens to induce cancer depends not only on the dosage and length of treatment, but even more on the tissue's inherent predisposition to malignancy."[93] To be sure, as Elmer Sevringhaus noted, estrogen was a "powerful stimulant to the growth of genital and mammary tissue. Any neoplasm may be expected to grow more rapidly under estrogenic therapy." The careful gynecologist first examined the patient thoroughly, but then "after adequate removal of a neoplasm there is no reason to deny the benefit of estrogen to a woman who had typical menopausal symptoms."[94] The leading medical textbooks, including the eminent "Cecil's," agreed: "The possibility of inducing cancer by large doses . . . cannot be disregarded," and "final judgment as to this must be reserved until more time has elapsed." The risks, however, were "remote" and "reports of cancer in the course of estrogenic treatment are exceedingly rare."[95]

Curiously, only postmenopausal women remained outside the reach of endocrinology. Gynecologists agreed that a "complex endocrine crisis" occurred at the moment of passage. But once it was completed, as Hamblen declared, "women are frequently happier and healthier than before. . . . Relieved of the anxieties of child-raising and the annoyances of menstrual function and reconciled to the cosmetic alterations of age, they acquire mental and physical vitality never before experienced and enjoy for a decade or two the best years of their life."[96] The postmenopausal woman, Novak insisted, was "graceful and serene," enjoying a "second flowering."[97] Even physicians who held a darker view did not recommend estrogen. Postmenopausal women did lose the markers of femininity. They "undergo a certain amount of masculinization at the climacteric," observed Hoskins. "This is manifested by increased hair growth on the face . . . a change in the pitch and timbre of the voice, and an increased angularity of bodily configuration."[98] In time, he continued, "the skin loses its elasticity and appears dry and wrinkled . . . and the features may assume the appearance of a masculine or neuter type."[99] Nevertheless, gynecologists, usually so active, were not ready to intervene.

Why the reluctance? Perhaps the inability to reach a consensus on whether these signs of aging should be thought of as diseases kept them from experimenting with lifelong administration of estrogen. Perhaps they thought it too expensive to put women on the drug for decades (although they could have cited insulin as a precedent). Or perhaps it reflected physicians' traditional definition of womanhood. So long as a woman was young and capable of reproduction, she was, medically speaking, interesting. But let her lose her generative capacity, become old and asexual, and she passed beyond gynecological, and male, purview.

THREE

Making Up, Making Over

THE FORCES OF COMMERCE AND CULTURE were as vital as the ambitions of science and medicine in promoting the use of hormones. Beginning in the 1920s, American drug manufacturers funded research on estrogen, allying themselves closely with investigators and physicians and opening a relationship that would become even more intimate over time. The companies also advertised and marketed their products with all the energy and assiduousness that young, profit-seeking companies could muster. They skillfully built on the images of women as they appeared in films, joining a message to make up with a message to make over. At the same time, American women proved to be remarkably receptive to these efforts. They welcomed the prospect of happiness, energy, and youth through pharmacology, ready to purchase the products that promised to deliver them.

I

Once investigators began to appreciate the influence of endocrinology on human biology, drug companies competed with one another to support research. They designed their own laboratories and funded investigations in medical schools and universities as well. As soon as a new drug appeared promising, albeit before its efficacy was known, they touted it to obstetricians, gynecologists, endocrinologists, and general practitioners through a variety of innovative marketing strategies. The pharmaceutical houses advertised in medical journals, sent out direct mail, and most important, dispatched an army of detail men, their sales force, to visit doctors in their offices.[1] Through these initiatives, they entered a new and mutually beneficial partnership with physicians. The companies became manufacturers of "ethical" drugs, and doctors, their exclusive dispensers.

Around the outbreak of World War I, American drug companies first began to think seriously about scientific research. German-made pharmaceuticals, which had dominated the market, suddenly fell into short supply, giving Americans the incentive to duplicate their manufacturing and research methods and create their own products. The companies brought the German model into their laboratories and purchased academic expertise. The university-based researchers were delighted to accept their funds, which otherwise were in scant supply. The federal government was not yet a major supporter of research (that had to await World War II and the subsequent growth of the National Institutes of Health); foundation grants were exceptionally competitive and not very munificent. "The drug industry," as one of its historians has noted, "represented a wealthy and untapped source of assistance."[2]

The relationship between industry and academy flourished. Between 1925 and 1930, E. R. Squibb increased its research fellowship support from $18,000 to $50,000 annually, a sum that represented 15 percent of its total research expenditures. By 1939, Merck was supporting seventeen research projects at eleven different institutions, hoping to enjoy the success that Eli Lilly had when it collaborated with the University of Toronto to create the first commercial (and highly profitable) preparation of insulin. Merck, along with other companies, hired university researchers to help design their own laboratories, recruit investigators, and identify promising areas to explore. For example, Merck appointed Alfred Newton Richards of the University of Pennsylvania as its key advisor, and the consultancy arrangement continued even as Richards brilliantly oversaw the government's World War II medical research effort.[3]

Just when chemistry introduced the possibility of more potent drugs and greater profits, regulatory efforts by medical organizations and government bodies made it advantageous for companies to assume a new identity as manufacturers of "ethical drugs." It was an odd use of the term—what made a drug ethical? But the answer was clear: unlike so-called patent medicines, ethical drugs were advertised only to doctors, and were available to patients only by a physician's prescription. Several considerations prompted drug companies to meet the requirements of ethical drugs. First, they believed it was in their best economic interests to ally with the medical profession. Earlier, when promoting patent medicines, the companies had irritated practitioners and professional medical organizations like the AMA. The point of conflict was not so much the dismal quality of the patent drugs but the fact that they were sold over the counter. From the doctor's vantage point, every patent medicine purchased represented one less office visit. Moreover, only ethical drugs could receive a Seal of Acceptance from the AMA's Council on Pharmacy and Chemistry, and, after 1929, that approval was necessary in order to advertise in the AMA journal. In effect, drug companies gave up over-the-counter sales in order to cement their relationship with doctors.

The companies were pushed in this direction by government regulations that followed on popular exposés of the shortcomings of patent medicines. The Pure Food and Drug Act of 1906 required companies to disclose the amount of alcohol and opiates in their preparations; then the Food, Drug, and Cosmetic Act of 1938 required them to list the drug's ingredients on the label along with adequate warnings against unsafe dosages. Any misrepresentations were punishable by law. Drugs now had to be registered with the Food and Drug Administration (FDA), and companies were required to demonstrate that they were safe for use as indicated. The regulations allowed one crucial exception: labels need not include the drug's ingredients and warnings if it was to be dispensed exclusively by physicians—ostensibly doctors would already know its contents and would be able to evaluate its side effects—and it was left to the companies to decide which drugs it would market for prescription only. Thus, the regulations shifted from informing consumers about the contents of all drugs to trusting physicians to protect their patients from harm. As Peter Temin notes, "the FDA had appointed doctors as the consumers' purchasing agents," transferring calculations of risks and benefits from consumers to doctors.[4]

All these considerations tilted the industry to the side of "ethical drugs." To the degree that they relied upon physicians' prescriptions, they did not have to spell out the content or dangers of a particular compound. They could charge higher prices for prescription drugs, since their potency and value were attested to by physicians. And what advantages they lost in secrecy they gained back in patent protection, enjoying exclusive rights to market a drug for a number of years. For its part, organized medicine was enthusiastic about the changes, not yet anxious about the intrusive effects of big government. To the contrary, by trying to do away with snake oil, the government was benefiting physicians who would now garner the financial rewards of more frequent patient visits.

Among the first drugs to carry the label of "ethical" were the new hormone preparations, which included insulin, thyroid, pituitary products, and the most important in terms of sales, estrogen. Because of its successful marketing of hormones, the house history of the German-based Schering Corporation calculated that "ethical sales in 1938, for the first time, exceeded sales of 'Saraka' [a home remedy for constipation] and marked Schering's transition to a primarily ethical operation. As an indicator of the importance to the company of the relatively new field of hormones, its output of these products increased about 500 percent over the next 10 years."[5] Eli Lilly underwent the same transformation through the sales of insulin. Once Lilly began mass-producing the drug for diabetics, under the trade name of Iletin, it "turned into a first-rank, research based pharmaceutical manufacturer. . . . The days of Creek Indian remedies were gone." And so were the days of hucksterism. "If every house can get a few good Specialities," one Lilly director observed, "we won't be so keen about cutting one another's throats on Fluid Extract of Spunkwater or Tincture of Chickweed."[6]

The results, just as the companies hoped, were greater consumer expenditures on drugs and very few complaints about higher prices. By the early 1930s, Americans were spending some $715 million annually on drugs, of which $190 million, or 26 percent, went to prescription drugs. Over-the-counter remedies (from Bromo-Seltzer and Sal Hepatica to Vicks VapoRub) still made up the bulk of purchases, but consumers increasingly were relying on physicians' prescriptions, and drug companies were increasingly dependent on prescription drugs for profits. None of these changes troubled health policy experts. "Even a year's supply of insulin for a victim of diabetes," reported one national committee exploring the costs of health care, "or the purchase of numerous bottles of 'tonic' for a 'run-down condition,' do not ordinarily involve costs comparable with those for a major surgical operation."[7]

Consumers generally bought drugs in small quantities and over time so that they did not feel a financial pinch. To be sure, the wealthier the family, the more drugs it bought, and the more urban the family, the more it bought. For example, New York in 1931 supported 13,000 registered pharmacists; California and Illinois, 8,000; Mississippi, 700; and Vermont, 380.[8] And some patients suffering from chronic illness, including hormone deficiencies, were unable to afford the drugs. But from the companies' perspective, cost did not seem to be the decisive factor in determining sales and they were happy not to compete with one another through competitive pricing. Were physicians enthusiastic about a drug and ready to dispense it, enough patients would find a way to pay for it.

The pivotal role of prescription drugs reinforced the companies' commitment to research and to partnering with physicians. By the 1930s, the drug companies had such a large stake in expanding the number of conditions for which physicians might prescribe their products that they readily sent them to any researcher who wanted to test a new use or to any practitioner who wanted to experiment with a few patients. They might well find benefits, publish a favorable case report, and thus provide good publicity and copy for advertisements. In turn, investigators' articles acknowledged drug company contributions, not to alert readers to a possible conflict of interest but to express gratitude.

Just how close drug companies had become to investigators is apparent in the advice they dispensed on the market value of particular research. For example, Gregory Pincus, who helped design the contraceptive pill, asked G. D. Searle, in 1941, whether the hormone derivatives that he had produced in his laboratory could be patented. Searle discouraged some of his efforts, encouraged others, and in general gave useful counsel. About one preparation that Pincus thought promising, it told him that a rival company had already made a claim that would "definitely prevent us from making that compound by any process and for any use. Our only course would be (1) to infringe and hope not to be caught, (2) to license . . . or (3) to show that the claim is invalid. The first would probably not be tolerated, and we are of the opinion that the other two are impossible, so it looks like a first-class stymie on that point."[9]

The record of drug company support of investigators and physicians warrants special emphasis not only because a now well-entrenched pattern had its origins in this period, but because these relationships brought the companies an entirely new degree of respectability within science and medicine. Their relationship to investigators and physicians was not one of supplier to dealer or wholesaler to retailer, but of collaborators in health care innovation and delivery. That the companies were seeking to maximize profit was overlooked by enough researchers and doctors to give a special cast to the drug company–medicine interactions. Trust and confidence, not caution and wariness, became the markers of attitudes and postures.

II

One principal result of this new relationship was the respect and even warmth with which practicing physicians greeted the representations and representatives of ethical drug companies. In the 1920s, they had expressed some degree of skepticism about company pronouncements. A prominent endocrinologist like R. G. Hoskins warned his colleagues about the inaccuracy of their claims: "The primary object of the . . . commercial propagandist is to sell merchandise. After all, the seductive remark, 'What do you care about science, Doctor, if you get results?' is an appeal to uncritical gullibility by which many a product of little worth has been foisted upon the medical profession."[10] His colleague Emil Novak was also disdainful of the "absurdities of the purely commercial literature with which physicians of the country are being deluged."[11]

By the 1930s, however, this skepticism was receding. The wariness with which physicians were likely to read an advertisement for a household product or greet a door-to-door salesman was often suspended when it came to the drug companies and their products. This was assuredly true for the mountains of printed materials that they received from the companies. One 1940 survey of physician attitudes found that 70 percent of them were "favorably disposed" to direct-mail drug advertisements.[12] A later and particularly sophisticated analysis of physician attitudes found, to the surveyors' genuine surprise, a deeply held regard among doctors for these advertisements, or what drug companies preferred to call "communications." From a carefully selected sample of 129 physicians, only three expressed strong negative feelings about the advertisements. Even more telling, "unsolicited remarks in favor of the large drug houses appear in 56% of the interview records; serious criticism in less than 5%."[13] Since drug companies were partners, not adversaries, doctors believed that the literature was accurate and trustworthy.

This confidence, once established, had remarkable resiliency. In the mid-1950s, the AMA, wanting to know whether its members considered pharmaceutical company promotions misleading or unethical, hired a polling firm to conduct some one thousand interviews. The findings pointed to an extraordinary amount of physician trust. A full 86 percent agreed that drug companies were "cooperating with the medical profession" very well or well enough. Almost every physician believed that drug company advertising made "an important contribution to the postgraduate education of doctors." What did physicians like least about drug companies? Any efforts to circumvent them and go directly to the public. Thus, drug companies ought never to give information about a new drug to a journalist before they give it to physicians. The survey also asked whether any physicians' practices with drug companies were unethical or unprofessional. Most doctors thought it proper to ask for extra drug samples for use by their families. They did, however, find it unethical to accept a gift from a drug company in return for preparing a professional paper and to sell patients drug samples that they had received free.[14]

Physicians' inability to perceive or acknowledge the commercial character of drug company activities shaped their personal relationships with drug salesmen. They greeted the detail man, so called because he reputedly knew every detail about a particular drug, with respect and even fondness. Although detail men had been calling on doctors since the 1890s, their importance, both to the companies and to the physicians, increased when drugs became ethical. The industry invested heavily in this sales force, fully aware that the intrinsic character of their assignment—visiting doctors during office hours—did not allow them to use their time efficiently. As a rule, the drug reps did not have to wait their turn with patients, but they could not expect an immediate audience, either. Paid no less than $1,250 a year and usually considerably more (which in the 1930s put them well into the middle class), they typically called on four to eight doctors a day.[15] They were not to sell drugs directly to the physician but to introduce and promote the company's product line. And because the detail man did not have an order slip in hand, he was better able to make the encounter seem something other than a salesman pitching a product to a gullible buyer.

Detail men were unusually successful in maintaining this fiction. Doctors on average received calls from five detail men a week; the great majority of them saw every detail man and consistently reported favorably on the meeting. Most physicians gave detail men ample time—usually around eleven minutes per encounter according to one survey. A remarkable 86 percent told interviewers that they were either "favorably disposed" or "enthusiastic" about detail men. When asked to explain their positive feelings, they emphasized detail men's education, exaggerating their scientific credentials and making them into bona fide medical instructors who could teach doctors about new products and supply additional information about old ones. "Sometimes," commented one physician, "they have to go back to their research laboratories to find what I want, but they always get it to me."[16] In the AMA's survey, two-thirds of physicians declared that "detail men are their most important source of product information"; about half said that their communications with detail men led them to prescribe a drug. The physicians also expressed a personal liking for the detail men, welcoming an "opportunity to relax between patients."[17] Roughly half characterized a good detail man as more a friend than a salesman, and two-thirds regarded them as more like investment counselors than insurance salesmen.[18]

For the detail man, winning the physician's trust was the necessary first step in selling a product. He was, first and last, a salesman, his attention riveted on the bottom line—the number of prescriptions written for his company's product. Just when the physician relaxed, the detail man moved in to get his job done. Drug company training sessions and "how to" books laid out the special skills and styles to woo a doctor customer.

Tom Jones, a detail man, listed the most crucial do's and don'ts in his 1940 primer, *Detailing the Physician: Sales Promotion by Personal Contact with the Medical and Allied Professions.*[19] Regularly invoking the maxims taught by Dale Carnegie, the book aimed to assist detail men "in making daily contacts productive so that sales may be influenced favorably." Jones himself had been so successful a salesman that he persuaded a doctor to write an introduction to the book, and the doctor stated, without embarrassment, that he had "listened attentively and patiently during thirty years of medical practice" to good detail men. How many other small businessmen, let alone company officers, would boast of having listened patiently to a salesman?

After advising would-be detail men on how to land the position, Jones suggested a variety of tactics for convincing a doctor to prescribe a company's drug. Although he urged the men to "really know your product" and to read *JAMA* regularly, he focused mostly on strategies of personal manipulation. His list of "Don'ts" in the doctor's office included: "Don't sit down. Don't overstay your welcome. Don't smoke. Don't fail to thank the doctor for his courtesy in granting you the interview. Don't be overly aggressive in recommending clinical uses—only make suggestions and never try to teach anyone how to practice medicine. When encountering doctor know-it-all, ask him questions; when you meet doctor snob, try to convey your expertise in the drugs you are selling. Don't sell medical students short—they are the future." Detail men who mastered these lessons would achieve success and an income "sufficient to provide most of life's comforts and a few of its luxuries."[20]

The growing power and profits of drug companies and their skillful use of detail men inevitably captured the attention of magazines like *Fortune*. Its August 1940 issue, for example, devoted a lengthy story to Abbott Laboratories. "Abbott Has a Way of Its Own," ran the banner. "Evidence: Sales and Profits Doubled in Five Years." The article opened with a fictional but creditable vignette of a detail man sitting in the waiting room of Dr. Brush (a graduate, no less, of the Columbia College of Physicians and Surgeons). After the detail man chatted up the nurse (complimenting her on her "big brown eyes"), sized up the patients ("not much charity work in the office"), and estimated the income of the practice ("moderately prosperous"), he took note of one particular woman. She was "middle-aged, nervous as a witch—well, it could be almost anything, but it wouldn't hurt to ask Brush if he'd seen that paper on 'The Effects of Estrogenic Substance on Anxiety in Menopause' and look for a chance to follow through on Abbott's Estrone and Estriol." Once ushered into the doctor's office, he found that the subject of sex hormones "came up without any effort; everybody is talking about them these days." His conversation then ranged from the research of Doisy to the ideal dose of estrogen ("because of the trickiness of the substance"). The detail man considered "Estrone . . . too expensive for routine sampling, but [he] resolved to wrangle a half-dozen 1-mg. ampules ($4.89 wholesale) for Dr. Brush if he could." As *Fortune* accurately noted, the detail man "is not known as a salesman. . . . He is not permitted to pursue to a vulgar clincher. He symbolizes the schizophrenic merchandising mechanism of the 'ethical' drug manufacturer."

The story went on to shrewdly analyze the quid pro quo in the drug company–doctor relationship, which was different from other business relationships. The drug houses gave up the ordinary modes of selling products "as a price for entering into a sort of partnership with the medical profession." They conducted their own research and also supported physician investigators to develop new drugs, and they assiduously kept physicians informed about new products and new uses for old ones. These efforts represented "a practical service to an astonishingly large number of practitioners who lack the time or inclination to read the scientific journals." In return, physicians prescribed the company's drugs, recommended them to colleagues, and warmly greeted the detail man. *Fortune* closed its article by translating all this into dollars: Abbott was enjoying record profits and had just joined the big five of drug company "ethicals."[21]

Taking advantage of their position as manufacturers of "ethical drugs," the companies also advertised extensively in medical journals, giving prominence to estrogen products. They competed for physicians' business by emphasizing the potency and purity of their products and the ease of administering them. Schering attempted to associate its estrogen product with good science by running an advertisement that included snapshots of the twenty-three researchers who had contributed to synthesizing the drug. Because of them, the text announced, "you can now successfully treat any disorder which may be due to the lack of . . . female sex hormone."[22]

Company promotions noted estrogen's value in treating sexual underdevelopment and amenorrhea but focused mostly on its use at the time of menopause. Since all women experienced the passage and at least some of them had painful symptoms, the potential market was sizable. Because the medium was a medical journal, there was no need to waste space describing menopause; gynecologists and endocrinologists were familiar with the female climacteric, did not doubt its clinical reality, and did not need a primer on hot flashes, sleeplessness, or depression. (As we shall see later, this was not the case with "male menopause" and the marketing of testosterone.) The advertisements also presumed that gynecologists and endocrinologists shared the cultural stereotypes of older women. Convinced that Americans prized youth and energy, companies sold estrogen not only on the basis of its ability to relieve menopausal discomfort but also on its ability to reinvigorate. The illustration accompanying Ciba's advertisement for its "Female Sex Hormones . . . For Therapy," was a photograph of five showgirls, shoulder to shoulder, inside an enlarged cosmetic cream jar, each poised to leap out, kick up her legs, and begin a snappy dance routine. Their hair was permed, flowered, and bowed; they were all bare to the shoulder, the quintessence of youth and glamour. The cream jar was not an accidental prop but an effort to have doctors and patients alike identify estrogen with makeup. In this same spirit, the image that Ciba inserted above the banner reading "Again the Winner" (which referred to its newest estrogen product) was of a beauty contest winner who amply filled her one-piece bathing suit. The message, latent but unmistakable: estrogen transformed weary menopausal women into beauty queens.

Other advertisements suggested that estrogen would make women less depressed and more feminine. Encouraging physicians to be attentive to women's moods, Schering's advertisements for Di-Ovocylin featured a young, attractive woman dressed in a flowing toga; leaning against a classical pillar, she gazed upward, her left arm extended, in a pose that at once suggested yearning and availability. The Harrower company also used a classical column as a prop. Its young female figure was in modern dress and stood upright, but she, too, looked into the distance, seemingly awaiting the message that the text provided: estrogen would bring her "untroubled calm." Still another advertisement featured a well-dressed woman and her teenage daughter walking out of a doctor's office under the slogan: "There Goes a Happy Woman."

The companies were not averse to negative advertising. In one presentation, a middle-aged woman, standing in a well-decorated living room, clasped her hands behind her and looked out glass doors into the distance, an abject air of depression and loneliness conveying the burdens of menopause. The text promised that "the patient approaching menopause can now be reassured. Correction of endocrine imbalance relieves those distressing symptoms which may accompany the 'change of life' period." Another manufacturer advertised its new estrogen product by featuring a row of mares in the background (their urine was one source for estrogen), machines in the middle ground, and then, in the foreground, five repeats of the same lined and grim face of a woman, looking so downcast as to evoke associations to a mental patient trapped in a snake pit. At least one Ciba advertisement promoted Di-Ovocylin by picturing a witch hovering over a bucket stirring her brew. The text proclaimed: "Gone are the days when it was thought necessary to consult a witch to cast out the female evil spirit . . . which possessed women during the menopause."

III

Drug company images and messages at once reflected and reinforced the readiness of American women to use a variety of products to enhance their appearance and performance. In part, their receptivity testified to the success of science and medicine in popularizing an image of the hormonal body as fluid and amenable to change. In part, too, it demonstrated the power of a celebrity culture to turn women into consumers. It was as if Johns Hopkins joined Hollywood to teach women how to use compounds, clothing, and cosmetics.

No one individual better exemplified these trends or did more to pro-mote them than the novelist Gertrude Atherton. In 1919, the then sixty-two-year-old Atherton received an offer from Samuel Goldwyn to join his Hollywood stable of "eminent authors." The invitation arrived at a time when Atherton was feeling especially restless and dissatisfied. Her novels had brought her a modicum of success, a group of dependable readers, and willing publishers, but not the reputation or money she thought she deserved. Moreover, as a woman who craved the attention and adulation of men, she was distressed by the unmistakable signs of age. When Goldwyn offered her a three-year contract with a promise of $10,000 for each script that became a film, she immediately accepted.[23]

Atherton arrived in Hollywood just when a fledgling film industry was beginning to produce star-studded extravaganzas, attracting audiences by providing enticing images of women made glamorous through the skillful application of cosmetics and costumes. Moviemakers presumed that the boundaries between image and reality were highly permeable and easily breached. Fashion and makeup could camouflage age and even gender. Fans proved them right, not only in terms of the number of the-ater tickets bought but also the consumer items purchased. Hollywood producers and publicists skillfully used the new genre of film magazines to encourage an identification with the stars. They crafted biographies of the leading ladies that were more fictional than real, emphasizing, for example, their humble origins (small town, store clerk) so that fans would identify with them. The film magazines, in turn, attracted adver-tisers eager to sell their products. As *Photoplay* boasted in 1920, avid moviegoers were "perfect consumers," enthralled by the glossy adver-tisements for the clothing and cosmetics that promised to make them alluring. "The movies," as one film historian observed, "primed them to buy and the movies focused their vision on what to buy."[24]

Atherton quickly recognized that no one was better at manipulating images than Samuel Goldwyn. He was a genius at marketing films and marketing himself. He changed his name and his persona several times, freely altering dates, including his birth date, to serve his ambitions. He did not settle on his name, and trademark, until he was almost forty years old. Indeed, the "Goldwyn" trademark became the subject of a dispute that ended up in court. Ruling in Goldwyn's favor in 1923,

Judge Learned Hand found that "a self-made man may prefer a self-made name."[25]

Atherton was initially intrigued by Goldwyn's skill at making the most vaporous illusion seem genuine. She described how carpenters built flimsy wood frames that appeared to be real houses on-screen and how seamstresses fabricated costumes that looked like authentic Scottish clothing. But her fascination with Hollywood's seemingly infinite capacity to make anything up was soon tempered by her own inability to make herself over. She still looked her age, and in Hollywood, more than elsewhere, she complained, an older woman was a "dethroned icon." Worse, the laurels she expected from her scripts did not arrive. Goldwyn rejected most of them and the only one he accepted was so totally altered by a nameless studio drudge that Atherton wanted no part of it. Her disillusionment and disappointment with Goldwyn and Hollywood survives in a group photograph of the "eminent authors." Atherton looks distinctly uncomfortable and out of place. Her hair is not fashionable coiffed, she is not made up, and a purse and umbrella dangle awkwardly from her hand. She deliberately averts her face from the camera. The other women smile broadly from under large-brimmed hats. Atherton sits there glumly, every bit the outsider.

Within a year, Atherton left Hollywood, settled in New York, and went back to writing fiction. But her stories, she conceded, were uninspired. Then one day she happened upon a newspaper article describing how a seventy-year-old Viennese surgeon, Dr. Konrad Lorenz, who had been suffering from "premature senescence," regained his former energy and stamina by undergoing rejuvenation at the hands of Dr. Eugen Steinach, director of the Biological Institute of Vienna. What captured Atherton's attention was a comment from Harry Benjamin, a New York endocrinologist, who had studied with Steinach. According to Benjamin, treatments were available for women to reverse premature senescence. "Women," he wrote, "were running to the Steinach clinic from all over Europe, among them Russian princesses who sold their jewels to pay for treatments . . . that might restore their exhausted energies and enable them to make a living after the jewels had given out."[26] Atherton immediately called Benjamin to arrange a consultation.

At her initial visit, as Atherton recalled in her autobiography, Benjamin examined her and found her to be "in perfect condition." Her lethargy, he explained, might be due to glandular deficiencies: "There should be a fresh release of hormones into the bloodstream." Benjamin recommended a course of "X-ray stimulation" treatments of the ovaries, eight in all, to be conducted over three weeks. Atherton immediately agreed. During the first two weeks, she grew even more tired, sleeping most of the time and worrying that she was "ruined for life." But then suddenly, "torpor vanished. My brain seemed sparkling with light. . . . I almost flung myself at my desk. I wrote steadily for four hours. . . . It all gushed out like a geyser that had been 'capped' down in the cellars of my mind, battling for release. That geyser never paused in its outpourings until the book was finished."[27] The only stops were to consult Benjamin about her health and to ensure that her depiction of rejuvenation procedures was accurate.

Benjamin published his own version of the case history in a New York medical journal. In "The Steinach Method As Applied to Women," she was Case Number 1. Although he did not name Atherton, he made little effort to disguise her identity. The patient's chief complaint was "mental sterility," with symptoms including "occasional sleeplessness . . . increasing lack of mental activity," along with waning powers of concentration and imagination. Benjamin instituted X-ray treatments to stimulate the ovaries; as the patient's hormone levels climbed, her creativity returned. He pronounced the case a complete success: the patient regained her vigor and, after treatment, "nothing could tire her any more."[28]

Atherton's best-selling novel *Black Oxen* took its inspiration from this medical encounter. Serving as a promotional tract for rejuvenation, it encouraged women to trust their bodies to their doctors. Atherton's archetype physician had nothing in common with the tyrannizing men that Charlotte Perkins Gilman described in "The Yellow Wall-Paper." Gilman's doctors had wanted to subjugate women; Atherton's sought to liberate them, to transform "sad and apathetic" patients into vital, energetic, and happy women. The book's plot was exceptionally clumsy and the shifts between reality and illusion altogether contrived. In fact, *Black Oxen* reads so poorly that the only possible explanation for its popularity had to be women's fascination with enhancement.

The novel's heroine, Mary Ogden Zattiany, was a wealthy, self-possessed American who married into European royalty. She first appears in the guise of a mysterious and beguiling figure attending opening night of a lackluster play. Her demeanor attracts the attention of a young newspaper reporter, Lee Clavering, but even as he falls under her spell, he is puzzled by her unusual mixture of youth and age. "In spite of its smooth white skin and rounded contours above an undamaged throat, it was, subtly, not a young face." Yet, no signs of aging were apparent: "She did not look a day over twenty-eight. There were no marks of dissipation on her face . . . but for its cold regularity she would have looked younger."[29]

In short order, Clavering becomes Mary's fiancé. He is the prize of a generational contest that the older woman, for once, wins. Clavering does suspect that Mary is not what she appears to be, but he cannot penetrate the mystery. Her more formidable antagonists are a small group of wealthy older women who had grown up with Mary; they do not recognize the rejuvenated creature but sense that she bears an uncanny resemblance to the Mary they once knew. For her part, Mary claims also to know their former friend, reporting that she is now recuperating in a Vienna sanitarium under the care of Dr. Steinbach (obviously Steinach). Steinbach "is a very great doctor. He will keep poor Mary's secret as long as she lives and nobody in Vienna would doubt his word."[30]

Midway through the novel, the new Mary lets the women in on her secret. She is the proof of medicine's ability to modify nature. They find the news disconcerting, convinced that rejuvenation is unnatural and indecent. "To have succeeded in making herself young again, while the rest of them were pursuing their unruffled way to the grave was a deliberate insult both to themselves and to God." As one of them insists: "Growing old always seemed to me a natural process that no arts or dodges could interrupt, and any attempt to arrest the processes of nature was an irreverent gesture in the face of Almighty God." Mary responds by describing how miserable she had been; lacking the stamina to cope with everyday life, she had entered a sanitarium. The chief physician explained that her endocrine glands "had undergone a natural process of exhaustion. . . . The slower functioning of the endocrines is coincident with the climacteric," a condition that was common and curable. Once her ovaries were stimulated through a series of X-ray treatments, her other glands would again function "at full strength and a certain rejuvenation ensue as a matter of course."[32] The physician urged her to undergo the treatment. "I was a promising subject, for examination proved that my organs were healthy, my arteries soft; and I was not yet sixty."[33] She hesitated, but when more conventional therapies proved useless and she became even more desperate to "regain my old will power and vitality," she decided "that science might be able to accomplish a miracle where centuries of woman's wit had failed."[34] The treatment's success exceeded everyone's expectations. "With my youth restored," Mary declared, "I have the world at my feet once more."[35]

Determined to make an irrefutable case for enhancement, Atherton has Mary insist that enhancement is indistinguishable from cure. To counter the argument that rejuvenation violates the rules of nature or God, Mary asks: "Why have a doctor when you are ill? Are not illnesses the act of God? They certainly are processes of nature. . . . When you are ill, you invoke the aid of science in the old way precisely as I did in the new one."[36] And she is confident that the future will bear her out. "The time will come when this treatment I have undergone will be so much a matter of course that it will cause no more discussion than going under the knife for cancer."[37] Borrowing a page from J. B. S. Haldane (whose book appeared the very same year), Mary declares: "Science has defeated nature at many points. The isolation of germs, the discovery of toxins and serums, the triumph over diseases that once wasted whole nations. . . . It is entirely logical, and no more marvelous, that science should be able to arrest senescence, put back the clock. The wonder is that it has not been done before."[38]

The women persisted in their objections that rejuvenation was "not only mysterious and terrifying but subtly indecent."[39] But Mary stood fast. Neither their contemptuous glances nor their phrases such as "ignoble vanity" and "abnormal renaissance" unsettled her.[40] "I do not merely *look* young again," she proclaimed: "I *am* young. I am not the years I have passed in this world. I am the age of the rejuvenated glands in my body." "Some day," she concluded in a paraphrase of Steinach, "we shall have the proverb: 'A man is as old as his endocrines.' "[41]

Black Oxen ends on this note of triumph, unreservedly defending and endorsing enhancement. But Atherton herself had to confront yet another expression of hostility, this one advanced by a group of ministers and women who found the novel "not fit reading for young people" and wanted it banned.[42] Several members of the New York State legislature proposed amendments to the state's obscenity laws to remove it from circulation. John Sumner, who had succeeded Anthony Comstock as the head of the New York Society of Vice, led the crusade, in alliance with the Woman's Christian Temperance Union. Not only had *Black Oxen* elevated science over religion (by promoting so unnatural a procedure) but it also championed gross immorality (through Mary's dalliance with a younger man). In fact, several cities, including Rochester, banned the book from their public libraries.

Atherton, like her fictional Mary, expressed her unrelieved scorn for her antagonists. The country was filled with "narrow-minded, ignorant . . . bigoted . . . puritanical . . . atavists who soothe their inferiority complex by barking their hatred at anything new. The very word Science is abhorrent to them, and, if they ruled the world, progress would cease."[43] She was fortunate to have the unflagging support of her publisher, the free-speech advocate Horace Liveright. "A censorship over literature," Liveright asserted with no little disdain, "is stupid, ignorant, and impudent, and is against the fundamental social principles of all intelligent Americans." In the end, the literary defense of science and medicine defeated censorship. Atherton and Liveright found an ally in the colorful Democratic minority leader, Jimmy Walker. Assuring his colleagues that "no woman was ever ruined by a book," he mustered enough support to defeat the measure.[44]

IV

Not content to let *Black Oxen* alone speak for her, Atherton went on the lecture circuit to promote the efficacy of and, no less, the morality of rejuvenation. She assured her audiences that sixty-year-old grandmothers would not become flappers. "Such a result is unheard of." Nor would they become vamps. "In the majority of cases . . . reactivation has nothing whatever to do with the renewal of sex activities." She also warranted that rejuvenation would not entail women conceiving more children. "What woman of sixty," she asked, "would be bothered again with the manifold demands of the reproductive cycle? She has done her duty. . . . What she desires is a life of her own, independent of men, asking nothing of them but friendship."[45] (In time, of course, in vitro fertilization would prove her wrong.) But the core of Atherton's message was that women should trust to medicine rather than submit to "the appalling dirty trick" that nature played by having them grow old. "We live in an age of scientific marvels, and those who do not take advantage of them are fools and deserve the worst that malignant Nature can inflict upon them."[46] The medical procedures, she promised, would benefit both mind and body. "Owing to increased, or renewed, glandular activity, which affects the whole body and, above all, the mental processes, the patient will very likely look many years younger."[47]

How did her readers, audiences, and women more generally respond to the message? To judge by the nearly one thousand letters that Atherton received—a biased sample to be sure but an intriguing one—it was an answer to their prayers. Women began writing to Atherton immediately after *Black Oxen* was published and continued to do so during the Great Depression, through World War II, and right up to her death in 1947. They were so drawn to the potential of medicine to reverse or prevent aging that they readily expressed their own fears and fantasies, and beseeched her for information on where to get the procedure and what it would cost. Taken together, the letters have a timeless quality, echoing hopes and expectations that are no less prevalent today.[48]

One common theme was a delight in the prospect of medicine being able to relieve the fatigue and exhaustion that burdened their lives. One writer thanked Atherton for being a preacher who "spread a gospel of hope to tired men and women."[49] Another was more fulsome: "I have, this minute, finished reading your very unusual book, *Black Oxen*. To say that I have been entranced and astounded is to put it mildly. My mind is in a whirl. That such a situation should arise even in fiction causes the wildest hopes. The renewal of one's strength! Has science actually done this or is it wholly fiction?"[50] Although she was already leading a busy life, she wanted the energy to do more. "I live for my work—*adore* it. I do a great deal of public work with these wonderful years. I get so tired,

I almost die. . . . If I can get a fresh hold—well that is a wonderful hope, at least. The hope alone has rejuvenated me about ten years, if nothing else comes from it."[51] In this same vein, a forty-five-year-old woman recounted that she always enjoyed "splendid health and plenty of vitality, but recently noticed that I tire easily and cannot bring the same enthusiasm to my work. If there is a chance of becoming rejuvenated, I should like to try it."[52]

Working women were hopeful that rejuvenation would give them an edge in a marketplace that demanded all-out aggression and competitiveness. "In as much as I am forty-nine and in a man's business," explained one woman, "it is terrifically important for me to retain as much youth as possible as long as I can, both mentally and physically."[53] Admonitions against tampering with the natural seemed insignificant compared to the chance to achieve economic success. "I am a business woman and have none of the inhibitions of the other kind of women. . . . Most of the woman with whom I have discussed it say: 'I believe in letting nature take its course,' but at least one of my acquaintances is going to take it. . . . So you have made two converts."[54]

One letter writer described with particular force the handicaps of age in an occupation in which appearance and demeanor trumped experience. "My only training has been along the lines of newspaper art in which field one must compete with very young people. To do this one must look and feel youthful—recover if possible the hopeful, enthusiastic energy of the twenties. It seems you have done this way past middle age and the fact that it has been publicized would seem to indicate that you are not necessarily desirous of keeping the method a secret. I am enclosing a stamped self-addressed envelope for your reply and trust that I may hear from you soon."[55]

For other women, fatigue and depression were intertwined. A single mother who thought she would be unable to afford the procedure offered to become a research subject in order to receive it; the risks were more than offset by the possibility that she might improve her life chances. "I am not going to burden you with the 'story of my life,' " she wrote Atherton. "But I can honestly say that it has been an uncommonly sad one. I fancy there are few people who are so utterly alone and in such difficult circumstances as my little son and myself. . . . Unfortunately, I am not able to afford it at present, but it has occurred to me that, as the treatment is in its early stages, they might be willing to have people to experiment upon, and so would give the treatment for a small sum."[56] Another single mother expressed this same desperation. "I am willing to make any sacrifice to regain my youth and vitality, for the sole purpose of being able to support my child. I am forty-one years of age. Work has been the chief ingredient of my life—a colorless drab existence. I married a physician twelve years ago. When my baby was one year old, I was discarded. I have no illusions and life seems very difficult at times. Can you please advise me where the Steinach treatment may be had?"[57]

Women also hoped to become more physically attractive, not to win beauty contests but to reduce social isolation. "I have a liberty bond and some money in the bank. How quickly I will use it if it is enough to pay for the treatment. . . . How can anyone resist who has the money? . . . I may be morbid but I can't bear to have anyone look at me and I am alone and need friends so much and want to be attractive, and I am not really old. If I can be made over . . . I shall thank you my dear Mrs. Atherton from the bottom of my heart."[58]

During the 1930s, economic and social dislocation exacerbated personal problems. "Not only has the Depression 'done things' to me," a fifty-year-old woman told Atherton, "but I have recently lost my husband . . . and it will be necessary for me to enter some field of business—but I cannot do so in my present 'let down' state. . . . I have read much on the subject of glands (all of them) as factors in our well-being and I am convinced if I could have the treatments you took I would be restored to a point where life would be worth living again. May I ask where you took your treatments . . . how long it takes, what is done, and the approximate costs?"[59]

All these women considered rejuvenation to be a thoroughly useful and legitimate enterprise. Since the goals were conventional and legitimate—physical, psychological, or professional improvement—the means to accomplish them had to be so as well. Their letters confirm the assessment that Carl Van Vechten, a literary critic and close friend of Atherton, offered about *Black Oxen:* "It is a book for flappers to laugh at, for middle-aged women to weep over, and for really aged ladies to be thankful for."[60]

FOUR

Controversies Forever

I N 1957, WILLIAM MASTERS, A YOUNG GYNECOLOGIST at Washington University School of Medicine in St. Louis, delivered a paper at the annual meeting of the American Gynecological Society that frankly intended to revolutionize the specialty. Citing findings from recent research, including his own, he proposed that decreased sex steroid production, essentially the loss of estrogen and progesterone, was responsible for the many physical and mental disabilities that aging women suffered. The consequences were system-wide, causing not only fatigue, depression, and loss of memory but also decreased bone formation and impaired cardiac functioning. Although the studies confirming these findings were few and small-scale, the negative effects of estrogen loss were so evident that gynecologists should administer hormone therapy to all postmenopausal women for life.[1] Conceding that "sex steroid replacement in the aged individual is an extremely controversial subject," Masters nevertheless insisted that it would counter "the general physical and mental disabilities associated with the aging process."[2]

Masters' paper was bold, a tour de force, laying the foundation for hormone therapy in the 1960s and '70s.[3] His mission was to make the care of aging women a central concern in gynecology. Gerontology should become a gynecological subspecialty, a goal that was almost as novel and controversial as estrogen therapy itself. Masters wanted his colleagues to abandon the traditional idea of a postmenopausal hormonal balance, to appreciate that the loss of estrogen put the aging woman into a state of endocrine imbalance that must be corrected. The logic of the case seemed so compelling and the benefits so dramatic as to outweigh potential risks or even the need to await more data.

I

Masters had already been investigating hormone replacement for almost a decade. In 1948, with his mentor Willard Allen (who together with George Corner had isolated progesterone), he published "Female Sex Hormone Replacement in the Aged Woman," a report on their effort "to find out whether or not some of the manifestations of aging are reversible."[4] The paper was preliminary and covered only thirteen women who ranged in age from sixty-four to eighty-two. Masters and Allen had given them a combination of estrogen and progesterone and then withdrawn it; bleeding followed immediately, and "the amount of bleeding closely approximated that which they had experienced in their active menstrual life."[5] In fact, right before the bleeding began, several subjects experienced the same "nervousness, irritability, and crying spells" that had years before preceded their menstrual periods. Because of this "close approximation," Masters and Allen hypothesized that through a combination of estrogen and progesterone, a "considerable rejuvenation of the reproductive organs can be brought about after atrophy has occurred."[6] They were not interested in the cosmetic aspects of withdrawal bleeding, which, as we have seen, had been induced by gynecologists in the 1930s in their young patients. Rather, they believed that the bleeding might provoke a "rejuvenation of a part of the reproductive tract," and, perhaps, exert a "beneficial effect on the body as a whole."[7]

Masters amplified this proposition in several subsequent papers, even as he admitted that his claims rested on thin data. He reported finding some regeneration of blood cells along the reproductive tract but none in the vagina and no restoration of functioning in the ovaries. "The complete burden of proof," he acknowledged, "remains, as it should, with the proponents of the concept of 'puberty to grave' support with sex steroids."[8] Still, he persisted, convinced that such a regimen would improve the lives of elderly women and, indeed, benefit the larger society. "One of the greatest public health problems of the present and future," Masters contended, "is the rapid increase in our aging population."[9] Treating aging as a disease and subjecting it to a potent, long-lasting, if expensive treatment was worth the investment.

Recognizing how little evidence supported hormone therapy, Masters not only inflated its promise but was not above resorting to emotionally loaded language and scare tactics to buttress his case. The aging woman was a victim of "steroid starvation" that turned her into a "neutral gender"; the "failing reproductive powers of the gonads" rendered her a "former female."[10] "With the approach of the 'neutral gender' age, the female pelvic organs shrink in size. The vaginal mucosa becomes atrophic, and the mons loses its full contour. The breasts sag, and girdle-type obesity becomes evident."[11] Surely a regimen that promised to protect women from such a fate warranted support.

Masters appreciated that he was proposing an all-out attack on the natural course of aging. As distinct from ordinary medical practice, he was seeking to move women from the normal (process of degeneration) to the optimal (state of well-being). For this very reason, he opened his 1957 paper with an unusual but crucial paragraph. "The only known member of the female gender to live past her period of reproductivity," he wrote, "is the human female. . . . The postmenopausal years represent, for her, a socially conditioned phenomenon." The extended longevity that women were now experiencing was in and of itself unnatural, the result of social, not biological, changes. The age of onset of menopause had not changed. Rather, "postmenopausal life had emerged as a consequence of the great increases in life expectancies developed over the last hundred years."[12] Medical and social changes had disturbed nature by extending the normal life span, and doctors, therefore, had to violate nature again to respond to the consequences of their own success. Since medicine had provided effective cures, it now owed women effective enhancements.

The goal announced, Masters mustered whatever arguments he could for using estrogen and progesterone. Absent data on actual outcomes, he had to rest his case on a general overview of female biology. He explained that while one purpose of ovarian functioning was to enable reproduction, a second and, from his perspective, "infinitely more important" purpose, was to serve as a "catalyst" through its hormonal production for many other vital physiological activities. "If only reproductive activity failed during the postmenopausal years, and sex steroid production were adequately maintained, there would be no definition of a 'neutral gender.' "[13] But the failure of the ovaries severely reduced the capacity of other glands to sustain their hormonal production. In effect, "the aging process slowly destroys the essence of interglandular stimulation." Put another way: "The postmenopausal inability of the gonadal elements to respond effectively . . . is the 'Achilles' heel' of the entire endocrine system." The resulting endocrine imbalance made women vulnerable to many degenerative diseases. The remedy was at hand: supplement "the clinically inadequate level of postmenopausal ovarian steroid production." Continued over the years, it would "recreate the proper balance pattern of endocrine secretory activity."[14]

The first disease on Masters' list was osteoporosis. Identified by Fuller Albright, a Harvard endocrinologist, in 1936, it was characterized by a decline in bone density that led to bone weakness and fractures. By the mid-1950s, endocrinologists had charted the complex ways that estrogen production facilitated calcium metabolism and storage, which were essential to healthy bones, and how declining estrogen levels in postmenopausal women impeded it. One research team, for example, using elderly women living in nursing homes as their subjects, first X-rayed them to determine bone density and then administered hormonal therapies over several years; because the women were permanent residents in the facilities, the team was able to repeat the X rays regularly. Although the number of subjects was small, the findings indicated an increase in bone density and, presumably, protection against fractures.[15]

The result struck Masters as altogether logical. Since sex steroids stimulated bone growth in young children hampered by slow development, he expected that they would have the same effect in older women; thus, "clinical osteoporosis immediately becomes a reversible process." It was no longer a natural or inevitable component of aging. "In essence," Masters concluded, "there is no such thing as old bone, if satisfactory sex steroid balance and adequate protein intake are maintained during the aging process."[16]

To further buttress the case for supplementation, Masters explored the relationship between hormone levels and cardiovascular disease. Although an understanding of the links among estrogen, cholesterol, and coronary artery disease was still rudimentary, a few epidemiological studies had documented a lower incidence of heart disease among premenopausal women compared to a matched sample of men. Estrogen had to promote cardiac health, for otherwise why were premenopausal women less prone to the disease than men? So too, why after menopause and "concomitant with decreased estrogen production" did women suffer a sharp rise in cardiac disease?[17] Investigators were also finding a rise in serum cholesterol levels in postmenopausal women and considerably more coronary atherosclerosis in women, regardless of age, whose ovaries had been surgically removed.[18] To Masters, the conclusion was unmistakable: ovarian hormones protected against cardiac disease.

Not content to stop there, Masters reported positive findings on sex steroids and cognitive functioning. Colleagues at the Department of Neuropsychiatry hypothesized that dementia might respond to hormone replacement therapy. "Such basic mental processes as memory for recent events, powers of definitive thinking, the ability to absorb new material," they observed, "have all shown significant improvement when a steroid-supported aged population was evaluated in a series of controlled experiments." To be sure, the improvement was not dramatic. Regardless of what dose was administered, a memory plateau was reached within a year and further treatment was futile. "There is obviously a level," they conceded, "of senile mental involution beyond which there is no recall under sex steroid influence." Still, hormonal therapy seemed to slow, if not reverse, dementia, and even limited improvement was valuable "when compared to the rapid disintegration rate of unsupported patients."[19]

Masters was aware that a small, vocal group of physicians were concerned that estrogen's stimulation of tissue growth, particularly in the uterus and breast, might increase the incidence of cancer. But he insisted that only indiscriminate and reckless use posed a problem. "Every precaution must be taken to protect her from the unpleasant results of uncontrolled therapy." He posited, again with no data, that if estrogenic compounds were used in cyclical fashion, stopping the therapy for one week each month, or if estrogen was balanced with progesterone, the risk of cancer would become negligible.[20]

Although Masters wanted all postmenopausal women on the therapy, he complained that the current products were too varied in composition and potency to make widespread use likely. "The market is flooded with innumerable steroid products which, when absorbed, vary greatly in strength. Even a physician experienced in steroid replacement techniques may easily err regarding the dosage schedule and functional strength of the product."[21] But he anticipated the development of an inexpensive and easy method to administer a product with a well-defined absorption rate to be given specifically to postmenopausal women. The physician would then have "the proper tools with which to prevent the inherent physiological waste accompanying the aging process."[22]

In the discussion at the American Gynecological Society meeting that followed Masters' presentation, some colleagues objected to the idea that the postmenopausal period represented pathology, and that older women necessarily turned into a neutral gender. They offered more benign images, insisting that the postmenopausal woman "may be just as intrinsically feminine as she was in her pre-menopausal years. . . . A beneficent nature has teleologically arranged by some kind of an unexplainable evolutionary process that the human female becomes non-reproductive . . . [at] middle age." In this way, she was spared the personal stress of reproduction and the larger society was spared the burden of coping with women having children at advanced age. The differences between Masters and his critics came down to an argument over nature—was it beneficent or cruel? Hanging in the balance was whether physicians should or should not give hormone supplements to all older women.

Masters did not yield an inch, even when a colleague accused him of being "a twentieth century Ponce de León who has not only searched for but possibly found a steroid fountain of youth." Masters remained convinced of the pathology of the postmenopausal period and defended his use of "neutral gender" on the grounds of public relations. "The term has demonstrated a drawing power far beyond our wildest expectation." We have been able "to direct the interest of the medial profession to what we believe is the greatest unexplored field in medicine today." Masters admitted to some uncertainty as to just when the therapy should begin, but he believed it "virtually impossible to overtreat."[23]

II

Through the 1960s and mid-1970s, a growing number of gynecologists endorsed and prescribed estrogen replacement therapy (ERT), most of them using estrogen alone, a few adding progesterone to it. In 1958, a little over a million and a half prescriptions were written for hormone therapy; by 1975, the number had climbed to 27 million.[24] The justifications generally followed Masters' position, but there were almost as many reasons offered as to why to give it and variations in what was actually given as there were doctors. Logic, a readiness to please, the blandishments of drug companies, and the demands of women all combined, despite an absence of data, to emphasize benefits and to relegate risks to the margins.

Gynecologists dispensed estrogen to relieve hot flashes and other symptoms associated with entering menopause. But what made ERT special, for doctors, for drug companies, and for women was its promise to counter the effects of growing older. Physicians prescribed it, and women took it, as an anti-aging tonic, an energy and mood booster. In the *Physicians' Desk Reference*, the book most often consulted by doctors for drug information, Ayerst Laboratories described Premarin, its best-selling preparation, as imparting "a gratifying 'sense of well-being.'" Its advertisements in medical journals for "estrogen replacement in the menopause . . . and later years" pictured a gray-haired, well-coiffed, fashionably dressed, and broadly smiling woman, engaging the attention of two older men. "Help Keep Her This Way" ran the caption.[25]

Tying estrogen to women's general well-being caught doctors' attention, and the media's attention. In fact, gynecologists partnered with editors of women's magazines to promote ERT, and the alliance was mutually beneficial. The physicians encouraged women to come to them for the drug and the magazines attracted readers through stories about retaining youth and beauty. *McCall's* magazine, in 1965, featured an article: "Pills to Keep Women Young: Eight Distinguished Doctors Who Have Done Important Work with Estrogen Therapy Tell of Its Exciting Results."[26] In the *Ladies' Home Journal*, Sherwin Kaufman, a well-established gynecologist, told "The Truth About Female Hormones," which was that a medical makeover was far superior to a cosmetic one. "It is perfectly natural for women," Kaufman wrote, "to wish to slow up the aging process and to remain more attractive. They don't hesitate to use contact lenses for failing eyesight, color rinses for drab-looking hair or caps for their teeth. Then why should they put up with the discomforts that afflict about half of them in middle age, when the menopause begins?" Eager to differentiate ERT from earlier quackery, he borrowed freely from Masters: "The truth is that menopause is a deficiency state which gradually defeminizes a woman physically. Estrogen replacement is not rejuvenation but an effort to extend the woman's own natural resources in line with her longer life span." Kaufman then offered yet another different type of distinction: "Treatment doesn't make a woman younger, but it does make her younger-looking." And he added a testimonial from one of his ERT patients, as though the case of one constituted data: " 'I just enjoy life more—I haven't felt so alive in years.' "[27]

However comfortable the relationship between women's magazines and gynecologists, it paled in comparison to the coziness of drug companies and gynecologists. Determined to market their estrogenic products and convinced that the number of customers for an anti-aging tonic was unlimited, the companies wooed doctors. The sales of ERT, they hoped, would rival oral contraceptives, the combination of estrogens and progestin that by 1965 was being used for birth control by some 6 million American women.[28] The goal was to increase the number of women using the pill and then, seamlessly, transition them to ERT. So first they would persuade women using oral contraceptives to avoid menopause altogether by shifting from the pill to ERT as menopause approached. Next they would persuade middle-aged women to begin ERT even before they entered menopause. Then they would convince postmenopausal women who were still resisting ERT that the aging process could be retarded or reversed. Were the campaign successful, all

American women would begin taking estrogen in the form of an oral contraceptive, remain on it through middle age, and then right before menopause (irrespective of symptoms) switch to ERT for the rest of their lives. Estrogen from adolescence through senescence.

Although the pharmaceutical houses sought testimonials from many physicians, they were especially keen to identify a lead spokesman, a gynecologist whose appearance, demeanor, and credentials would assure women that a daily dose of estrogen would make them happier, healthier, and more youthful. They found their man in Robert A. Wilson. Wilson's medical credentials (more than forty years in private practice as a Brooklyn gynecologist), his appearance and demeanor (gray hair, confident), his use of ERT (he had prescribed large doses to patients for decades), and his unyielding belief in its potential to prevent aging and disease appealed to the companies. In 1964, Wilson received $31,350, purportedly as a "research grant," from three pharmaceutical companies—Ayerst Laboratories (Premarin), the Searle Foundation (Enovid), and the Upjohn Company (Provera).[29] The funds were to enable him to establish a foundation that would promote the anti-aging properties of estrogen. In short order, the Wilson Research Foundation became a publicity machine and clearinghouse. It distributed pamphlets, arranged for medical lectures, and provided doctors with information on hormonal therapy. The foundation also developed, maintained, and circulated a nationwide list of physicians who prescribed the products. Physicians who put their name on the foundation list were soon deluged with referrals.

Wilson's willingness to serve as point man dramatically transformed his career and fortune. Between April 1963 and August 1964, he, who did not hold an academic position and was not an experienced clinical investigator, published thirteen articles in major medical journals, all of them spelling out the benefits of estrogen replacement therapy and its anti-aging qualities. As a pundit in the popular press, he assured readers of *Look* and *Vogue* that ERT would prevent twenty-six symptoms of menopause, including hot flashes, melancholia, loss of memory, backaches, and anxiety.[30] Even more, estrogen would make women "romantic, desirable, vibrant." They would "grow *visibly younger* day by day until they are transformed into the exciting vibrant females they were before the 'change.' "[31] All they had to do was to telephone the Wilson Foundation and obtain the name of a physician who prescribed ERT.

Wilson more than fulfilled drug company ambitions when his 1966 book appeared under the inspired title *Feminine Forever*. During its first six months in print, it sold over 100,000 copies, and within a year, it was available to readers in seventeen countries.[32] (Seeing they had a winner, the companies also paid Wilson to lecture around the country.) *Feminine Forever* had an extensive bibliography, as though it had been thoroughly researched. In fact, it was a collection of personal observations and anecdotes. Over the course of forty years, Wilson recounted, he had administered estrogen to over 5,000 women and "without exception, every case I treated . . . showed some degree of improvement. In many cases total avoidance of all menopausal symptoms was achieved, and the percentage of marked amelioration was surprisingly high. In the entire realm of medicine, there are few forms of therapy with a more consistent record of beneficence."[33]

Estrogen was no less potent against the ravages of aging. "Estrogen therapy," Wilson insisted, "is a proven, effective means of restoring the normal balance of her bodily and psychic functions throughout her prolonged life. It is nothing less than the method by which a woman can remain feminine forever."[34] Like Masters and Atherton, Wilson described menopause as nature's dirty trick, robbing women of their womanhood. When after menopause a woman's ovaries "shrivel up and die," she becomes the "equivalent of a eunuch. . . . The entire genital system dries up. The breasts become flabby and shrink, and the vagina becomes stiff and unyielding. The brittleness often causes chronic inflammation and skin cracks that become infected and make sexual intercourse impossible."[35] Shrivel, shrink, brittle, flabby—these were the marks of natural aging. Why suffer them when a remedy was at hand? Indeed, a woman had the patriotic duty not to suffer them. "Multiplied by millions, she is a focus of bitterness and discontent in the whole fabric of our civilization."[36]

What then should women do to assert their "right to remain women"?[37] Everything came back to ERT. "Estrogen therapy doesn't *change* a woman," Wilson declared. "It *keeps her from changing*, that is, from suffering 'living decay.' "[38] To watch a "pleasant energetic woman" turn into "a dull-minded but sharp-tongued caricature of her former self is one of the saddest of human spectacles."[39] With ERT, women did not have to "live as sexual neuters for half their lives."[40] Moreover, "women rich in estrogen tend to have a certain mental vigor that gives them self-confidence, a sense of mastery over their destiny, the ability to think out problems effectively, resistance to mental and physical fatigue, and emotional self-control."[41] In effect, ERT bested nature.

Wilson assured his audiences that estrogen was no aphrodisiac. "No woman, I repeat, is in danger of losing her sexual self-control due to estrogen therapy. Because it invigorates the female sex organs and the total personality, estrogen tends to make sex more enjoyable—regardless of age. But it does not stimulate sexual aggressiveness."[42] The message may well have been aimed at husbands as well as wives, assuring both of them that ERT would not upset whatever balances had been struck within the marriage. The women would be Feminine not Masculine, Forever.

Just what constituted the ERT regimen that Wilson was promoting? Most women who followed it took Premarin, the Ayerst Laboratories product that was made from pregnant mare urine (hence the name), which by 1966 had become one of the leading prescribed drugs in the United States. But there the uniformity stopped. Recommendations on proper dosage came in a bewildering variety and physicians often made up their own particular regimens. The drug companies facilitated and encouraged variations. To "permit flexibility in dosage adjustment," Ayerst, for example, offered Premarin in four different preparations, ranging from 0.3 to 2.5 milligrams. It suggested that physicians start menopausal and postmenopausal women on 1.25 milligrams daily and then increase or decrease the dose as they saw fit.[43] Gynecological texts also encouraged improvisation. "The physician," one suggested, "should regulate the dosage by the degree to which hot flushes are relieved, insomnia corrected, and a feeling of 'well-being' produced."[44] Other texts agreed. "As there is a great individual variation in response to estrogen therapy, no scheme can be regarded as infallible, but must be used only as a norm from which to deviate."[45] Some physicians even allowed the patient set the dose, since only she knew when she had achieved "a sense of well-being."

One survey of gynecological practices highlighted the exceptional variations in estrogen regimens. In 1973, Helen Jern, a young New York gynecologist whose 1,500 postmenopausal patients typically came through Wilson Foundation referrals, conducted an informal survey of prescribing patterns.[46] "Most postmenopausal patients," she found, "are treated with estrogen alone, the majority with a conjugated estrogen, often Premarin." The most popular dose was the one recommended by Ayerst—1.25 milligrams—daily. Some gynecologists, however, reduced it to every other day if breakthrough bleeding occurred; others lowered it gradually to 0.625 and 0.3 every other day.[47] But there was no conclusive data on whether 0.625 every other day was more effective than 1.25 daily or whether 0.3 milligram every other day was even a therapeutic dose. Wilson himself followed a grab bag of practices. He usually began his patients on a 1.25 milligram tablet of Premarin daily, but instead of reducing the dose over time, he moved off the twenty-eight-day cycle. Women who were ten years past menopause went to a forty-two-day cycle: 1.25 milligram tablet of Premarin for the first thirty days, then a "10 milligram progestin tablet . . . through the forty-second day." On the forty-second day, Wilson told his patients, "both tablets are discontinued, and two or three days later—as a token of your restored femininity—you will menstruate once again. Then you resume taking estrogen tablets on the fifth day of menstruation." After his patients followed this regimen for three to five years, Wilson lengthened the cycle to fifty-two days and, around the six-to-eight-year mark, increased it to sixty-two days.[48] He never explained how he arrived at these numbers beyond saying it was based on forty years of experience. For very elderly women just beginning ERT, Wilson sometimes prescribed very large doses. "You are long past menopause," he told them, "and in serious mental and physical difficulties. To name just a few aspects of the total syndrome: deep untractable depression bordering on melancholia, crippling osteoporosis (brittleness of the bones), cholesterol content of the blood so high as to produce imminent danger of strokes or heart attack."[49] Accordingly, he gave them a "crash program," such as doubling the doses of Premarin (2.5 milligrams) for twenty-eight days and the dose of Provera (progestin 20 milligrams). Crash programs were especially suitable for an elderly woman about to marry "a healthy, vigorous man" and needed help to "fulfill her wifely obligations."[50]

Since gynecologists considered psychological depression a prominent symptom of menopause, they also prescribed sedatives or more potent tranquilizers along with estrogen.[51] Jern, for example, gave 5 milligrams of Valium to women beginning hormonal therapy. "It calms the patient, affords better sleep, and paves the way for a quicker approach to the time when, under the influence of estrogen, the nervousness, anxiety and fears disappear."[52] Some pharmaceutical companies marketed a combination sedative-estrogen. Milprem, the widely used product of Wallace Laboratories, was composed of estrogen and Miltown. ("Change of life . . . requires more than just reassurance.") Menrium, which was produced by Roche, combined estrogen with Librium. ("Because the menopausal syndrome can be as much an emotional as a hormonal problem.")[53]

III

If patients could not be certain precisely what they were taking, they did know why they were taking it. Gynecologists frequently complained among themselves that they were being bullied by their patients for the preparations, and swapped stories about demanding women. "They've read Dr. Wilson's book and all of the magazine articles about estrogen therapy," one Miami gynecologist recounted, "and they insist that I give them the pills. When I don't, they accuse me of being a 'medical reactionary' and worse." Another reported that a twenty-five-year-old woman requested an estrogen replacement prescription. When he suggested she return in twenty years, "she walked out in a huff." Sherwin Kaufman, who became a medical celebrity by promoting estrogen, believed that "the situation has gotten ridiculous. Women come in asking for 'the youth pill,' and they say, 'check my estrogen level.' From what they've read they think it's as simple as driving into a gasoline station and having their oil checked."[54] With medical information now circulating more extensively than ever before—courtesy of Wilson, other physicians, and the women's magazines—patients were starting to tell doctors what they wanted. Indeed, Wilson recommended that women select their doctors not on the basis of their training and skill but on their readiness to make them happy by prescribing ERT. "It is not necessary to seek out a specialist," he counseled. "Any medical practitioner can treat you and keep you under the necessary observation. His main qualification, other than general medical competence, should be a genuine interest in your health and happiness; in short, the ability to sympathize with the needs and problems of his patients."[55] In this way, estrogen regimens both reinforced traditional ideas about woman's proper place and began to empower women against their doctors.

By all accounts, ERT was particularly favored by upper-middle-class and educated women, so much so that retrospective analyses of the efficacy of estrogen (comparing past health outcomes for women on ERT with women who were not) were confounded by the fact that the most health-conscious and financially comfortable women were on the regimen. The appeal to appearance and greater mental and physical energy obviously fit well with their lifestyle and ambitions. But ERT also crossed class lines. The Wilson Foundation, for example, referred women in New York who could not afford a private physician to Bellevue Hospital's menopausal clinic, where they often became the patients of Helen Jern. And Jern, curious about the impact of estrogen on their lives, sent her patients a questionnaire and included some of their responses in her 1973 book, *Hormone Therapy of the Menopause and Aging*. Since Jern was a proponent of ERT, her choice of what to quote was obviously biased. Still, the statements ring true, echoing what women had written Gertrude Atherton several decades earlier.

They were not, as Jern described them, simple cases. The women had long histories of physical and psychological complaints, ranging from extreme fatigue to alcoholism. But soon after beginning hormonal treatment, they reported that their debilities diminished and, in some cases, disappeared. As long as they remained on ERT, they described themselves as more calm and confident. Take the case of Ms. A.L., a forty-year-old divorced single mother. "In 1965," she wrote Jern, "I read a story in a magazine about how Dr. Wilson had helped many women who were suffering from symptoms similar to mine." She called the Wilson Foundation and was given Jern's name. At her first visit, she complained of "extreme fatigue . . . depression, arthralgia of the spine, dryness and aging of the skin." She was still menstruating regularly but her periods were "scanty." Jern gave Ms. A.L. an injection of calcium, some vitamin B, and iron, and also 1.25 milligrams of Premarin and 5 milligrams of Provera. Her response was apparently remarkable. "After two-and-one-half weeks, I awoke one morning with a sense of well-being I'd never experienced before. . . . It was the beginning of a new life for me. . . . I had the courage to take a job I'd always wanted. . . . My state of mind is happy and I'm terribly excited about life and all that I have been able to accomplish because of my new health." She remained on it for seven years, insisting that it "has given me the chance to enjoy the second half of my life, to establish a fine relationship with my son, the man I love, and friends."[56]

Other working-class women reported restored youth and energy. One unmarried woman consulted Jern ten years after menopause, when she was fifty-seven. Her complaints were hot flashes, inability to concentrate, and general fatigue. Although she should have been working full-time to support herself, she could manage to hold only a part-time job. "The doctors called it 'nervous anxiety,' " she told Jern, "I called it a living death." Jern put her on Premarin and Provera, with apparently great results. "I started the return of my menstrual periods, which was my cure. After reestablishing my periods, my flushes and fatigue gradually disappeared." Later, when her symptoms returned, Jern gave her higher doses of Premarin and the woman declared herself "again normal and fine. I have a feeling of general well-being, and I am returned to youthfulness and optimism." Jern, of course, was delighted. "This patient was feeling completely well and looking 15 years younger than her age."[57]

A fifty-eight-year-old mother of seven children was another satisfied patient. Jern saw her some seven years after menopause, with complaints of fatigue, nervousness, and inability to concentrate. Jern prescribed 1.25 milligrams of Premarin with 10 milligrams of Provera, but when that proved ineffective, increased the dose to 1.85 milligrams of Premarin. "Estrogen-Progesterone therapy," the woman declared, "revolutionized my entire life. . . . My husband came back to me . . . I look forward to life with a zest which I never had before." Her husband concurred: "I could see the improvement in her health, her vitality, and in her sexual response." It took twenty years off her age and "saved and restored our marriage."[58]

Even much older women benefited from the regimen. A sixty-nine-year-old who had undergone menopause twenty-four years earlier arrived at Jern's clinic complaining of dizziness and fatigue that often left her too weak to eat or get out of bed. Jern prescribed 1.25 milligrams of Premarin and 20 milligrams of Provera. (Jern's dosing varied greatly, too, but she never explained why.) Within two weeks, the patient improved dramatically. "I gained weight, sleep better and have energy." Two years later she informed Jern: "It is a miracle to find myself at the age of 71 enjoying better health after several years of misery. I am stronger, my vision has improved and I have experienced an orgasm."[59]

Many of Jern's patients saw their renewed bleeding as the return of menstrual periods, and thus tangible evidence of their youth restored. (The association of blood and youth goes deep. Rosalie, the heroine of Thomas Mann's 1953 novel, *The Black Swan*, was distressed by the onset of menopause and, longing for romance, became infatuated with a much younger man. No sooner did he respond to her than she experienced a sudden bout of bleeding, which she interpreted as menstruation and a sign of returning youth. In fact, the bleeding came from a fatal uterine tumor.) The women were encouraged in this confusion by their doctors. Gynecologists regularly referred to the bleeding as a "token" of their restored femininity, a sign that "you will menstruate again." The doctors really knew better. Menstruation in premenopausal women came at the close of a "fertile cycle"; cyclical bleeding in postmenopausal women marked the termination of a planned "sterile cycle." But by not explaining the distinction, they practiced the same cosmetic endocrinology with older women that they had practiced with younger ones. If cyclical bleeding created an illusion of restored youth, they would not shatter it.

IV

There were dissenters to ERT who were deeply concerned about possible side effects. The difficulty was that their negative views rarely circulated outside narrow medical circles. Women's magazines were not lining up to interview them and medical journals were in no rush to publish negative results. The most telling criticism was of ERT as an anti-aging tonic. Edmund Novak, son of Emil, and also a member of Johns Hopkins Ob-Gyn department, was among the most vehement. He had taken over both his father's practice and his father's highly regarded *Novak's Textbook of Gynecology*. Writing in the *Johns Hopkins Medical Journal* (a publication unknown to non-physicians and barely known to physicians), Edmund Novak was appalled that *Feminine Forever* was influencing gynecological practice. The "wild claims" of the Wilson school, he insisted, lacked scientific support, discredited gynecology, and misled women. "We deplore the unfortunate publicity which seems to promise middle-aged women that hormones can reverse the inexorable inroads of age. Let us all be assured that endocrine agents are not a panacea for the aging process."[60] Novak, himself, was persuaded that the postmenopausal state was not pathological and that elderly women did not need potent drugs. To be sure, the moment of passage into menopause represented an endocrine imbalance, but later, a natural readjustment occurred. Novak estimated that about 15 percent of women had symptoms severe enough "to warrant endocrine therapy if mild sedatives and tranquilizers do not avail." Even so, physicians should "advise these patients that treatment will be only temporary, and that even if none were given, the symptoms would ultimately disappear."[61]

Novak certainly did not want to see endocrine therapy become a long-term treatment, unpersuaded of its ability to prevent atherosclerosis or osteoporosis. When he and a team of Johns Hopkins colleagues conducted autopsies on eighty-five women whose ovaries had been removed, and then compared them to controls, they found no evidence of increased cardiac disease. "We are not discounting the possible guarding effect of estrogen," Novak declared, "but feel it may be only one of many factors involved."[62] Noting that some gynecologists used "extremely high doses of estrogen if there should be any hope of preventing progression of the osteoporosis," Novak found the supporting evidence insufficient.[63]

Finally, Novak rejected the association between restored menstruation and restored youth. Gynecologists knew perfectly well that "menstrual activity" was not menstruation. "What 60-year-old woman," Novak caustically asked, "needs continued menstruation as a 'badge of femininity?' "[64] ERT would not turn back or slow down the clock. "Hormones must not be regarded as a panacea to the inevitable aging process."[65] His overall conclusion was negative and without qualification. "Until there is definite evidence to the effect that estrogen is an absolute barrier to cardiovascular disease, and that every castrate woman of necessity suffers from estrogen deficiency, then our treatment will not include routine replacement."[66]

Opponents of ERT were even more worried about potential risks, especially the risk of cancer. The data had become sufficiently compelling to put some pro-ERT proponents, albeit not all and certainly not Wilson, on the defensive. As early as 1937, clinical investigators had correlated long-term, high-dose estrogen to cancers in laboratory animals.[67] The finding set off a debate that continued in the 1960s, as gynecologists administered higher doses of estrogen over longer periods to their patients.[68] Wilson, predictably, dismissed the "myth of cancer." "The truth is exactly the opposite. There is increasing evidence that estrogen has a preventive effect on breast and genital cancers." His major source was a 1962 retrospective study in which 300 women who used ERT were reported to be free of cancer when "given normal odds, as established by medical statistics, eighteen cases of cancer—either of the breast or the uterus—would normally be expected in this group." Ignoring all the statistical and sampling problems, Wilson concluded: "There is evidence that women who stay estrogen-rich throughout their lives will remain happily cancer-poor."[69]

Other ERT proponents, however, worried that too high or too prolonged a use of estrogen would overstimulate tissue in the uterus and the breast, and might produce cancer. Ayerst, without openly acknowledging the possibility, cautioned against the uninterrupted use of Premarin. "To avoid continuous stimulation of the breast and uterus, cyclical therapy is recommended (3-week regimen with 1-week rest period—withdrawal bleeding may occur during this 1-week rest period)."[70] Some physicians believed even an interrupted estrogen regimen carried risks, noting that gynecologists who performed biopsies of cervical tissue often found an abnormal proliferation of cells in the endometrium (the mucous membrane lining of the uterus). As enthusiastic as Jern was about ERT, she, too, conceded that long-term use carried risks. "It is established that such administration of un-opposed estrogen may produce endometrial hyperplasia and polyps and hyperplasia of the breasts. There is always the potential of further histological change which may eventually lead to the development of carcinoma."[71]

Despite these concerns Jern, and many others, kept patients on hor-monal therapies for years so that they could have "a new influx of energy, a revitalization of ambition, a return to a youthful positive atti-tude . . . and a desire to be a useful and productive person."[72] Achieving such results required that patients remain on ERT for extended periods of time—benefits quickly disappeared as soon as treatment stopped. To justify her practice Jern quoted a statement from a sixty-six-year-old patient: "My feeling is that when a woman has to earn her own living it is desirable to do whatever is possible to retard the aging process so that living and working may be faced with optimism and pleasure rather than live in a state of depression, desperation, fear, and misery. I finally decided I'd rather take my chances with hormone therapy. At least living through old age would be pleasant, even if it did end with a cancer."[73]

Other ERT proponents also continued dispensing ERT, disregarding the risk of cancer. "There have been a few relatively small surveys of the incidence of uterine (and breast) malignancies during prolonged estro-gen therapy," conceded Sherwin Kaufman. But to rule in such an associ-ation required "much broader studies involving much larger numbers of women and carefully documented data over a great many years."[74] Until these studies were completed, Kaufman would administer ERT. That he was setting a far more rigorous standard for establishing the risks of estrogen than the benefits of estrogen apparently escaped him.

To Novak and his allies, continuing estrogen therapy while awaiting further research represented "a short-sighted attempt to justify their policy to keep women 'endocrine-rich and cancer-poor.' "[75] The "possi-bility that estrogen over a long period of time might be an adjuvant fac-tor in the evolution of endometrial cancer" was too strong to condone such practices. Even were estrogen not the "sole causative agent," the very fact that "many gynecologists such as ourselves feel it may play a very important role" should prompt colleagues to err on the side of caution.[76] Because pathologists were "impressed by the large number of postmenopausal endometria in which extreme degrees of hyperplasia and even endometrial adenocarcinoma are observed in women who have a history of prolonged estrogen therapy," ordinary practitioners should change their prescribing habits.[77] This message, however, went unheeded.

The cancer-ERT controversy became even more intense when findings linked the use of DES in pregnant women (usually prescribed to prevent miscarriages) to subsequent cancer in their children. In 1971, Arthur Herbst, a gynecologist at Harvard Medical School, reported very strong associations between DES use and a rare form of cancer, adenocarcinoma of the vagina. Over a three-year-period, Herbst diagnosed this highly unusual malignancy in seven women between the ages of fifteen and twenty-two. The cluster of cases astonished him. When he noted that all the women had been born between 1946 and 1951 in New England hospitals, he did some medical detective work and found that all of their mothers had taken DES during the first trimester of their pregnancy. "Maternal ingestion of stilbestrol during early pregnancy," Herbst announced, "appears to have enhanced the risk of vaginal adenocarcinoma developing years later in the offspring exposed."[78]

That same year, Peter Greenwald, an epidemiologist and director of the New York State Cancer Control Bureau, published a similar finding. Examining the New York State Cancer Registry between 1950 and 1970, Greenwald found five cases of adenocarcinoma of the vagina in women under thirty. It turned out that four of their mothers had taken DES during pregnancy, the fifth, another synthetic estrogen product.[79] Based on Herbst's and Greenwald's findings, the New York State commissioner of health, Hollis Ingraham, sent letters to all New York State physicians warning them about the danger of administering DES to pregnant women. In short order, the FDA issued an alert advising physicians against the use of DES in pregnant women.[80]

The findings on cancer did not stop with DES. In 1975, two new studies published in the *New England Journal of Medicine* linked estrone, one of the two principal ingredients of Premarin, to the rising incidence of endometrial cancer in postmenopausal women. In the first, Harry Ziel and William Finkle of the Kaiser Permanente Medical Center in Los Angeles analyzed the case histories of ninety-four Kaiser patients diagnosed with endometrial cancer to learn that 57 percent had taken Premarin compared to 15 percent of matched controls.[81] In the second, Donald Smith and colleagues from the University of Washington also found an association between women on ERT and endometrial cancer. Examining the medical records of 317 women diagnosed with the disease between 1960 and 1972 at their medical center, the team found that "the risk of endometrial cancer was 4.5 times greater among women exposed to estrogen therapy." The finding was especially troubling because it revealed "a pattern of endometrial carcinoma developing in large numbers of persons who do not possess the previously reported constitutional physiologic features associated with the disease."[82]

The publication of the two articles brought a new level of scrutiny to ERT, with just about every shade of opinion represented. The *NEJM* itself commissioned two editorials to comment on the papers, and both staked out a middle-of-the-road position—a risk of cancer existed but it was minimal. "The data reported in the two articles"—maintained Kenneth Ryan, a gynecologist at the Boston Hospital for Women—"indicate a risk of endometrial cancer increased five to 14 times in women taking estrogen." But rather than draw his conclusions from a comparison to a normal population, he emphasized instead that a 3- to 9-fold increased risk of endometrial cancer occurred in obese women and a 17-fold increased risk of death from lung cancer in pack-a-day smokers. These comparisons, he explained, "are testimony to the hypothesis that one might take estrogens for valid medical indications with a potential risk of cancer comparable to the self-abuse of overeating or smoking."[83] His reasoning was ever so odd, but others invoked it as well. Why compare estrogen-using women with substance abusers of food and nicotine? After all, physicians were prescribing estrogen; nicotine abusers were buying their own cigarettes, and overeaters, their own junk food. And what relevance did the risk of obesity have to the risk of estrogen? The second editorial, by Noel Weiss, a cancer epidemiologist, conceded that the probability of endometrial cancer occurring in non-estrogen-using postmenopausal woman was only 1 per 1,000, but among estrogen users, it was 4 to 8 per 1,000, a statistically significant increase in risk. But since the data were retrospective, gathered from epidemiological studies, and not prospective, drawn from clinical trials, Weiss remained agnostic. "The data available at present simply do not permit an answer that carries much conviction. . . . Despite the urgent need for answers, there is little choice but to remain in the dark for a few years more."[84] What should be done in the meantime? For both Weiss and Ryan, physicians should monitor their patients on ERT very closely and drug companies should devise a safer product.

Others, particularly drawn from the newly emerging women's and consumers' organizations, took a far more aggressive anti-estrogen position.[85] Feminists published articles with titles like: "Promise Her Anything But Give Her . . . Cancer."[86] Or: "Taking Men Out of Menopause." *The Ms. Guide to a Woman's Health* in 1981 unequivocally declared that "Estrogen replacement therapy (ERT) is a dangerously overused treatment. Avoid it if at all possible."[87] Menopause was a normal part of the life cycle that should not be medicalized. The cancer threat also led Consumers Union and Ralph Nader's Citizens Watch to flatly advise against ERT.

The Food and Drug Administration, which had a long record as a risk-averse regulatory body, also came out against ERT. "Estrogens are valuable drugs, with many beneficial medical purposes," the FDA commissioner Donald Kennedy declared. "But we at the FDA are increasingly concerned that estrogens are used too frequently and for too long, especially since their use for extended periods is associated with a 5 to 10 times higher risk of cancer of the uterus."[88] The risks were all the more unacceptable because estrogens were "given to otherwise healthy women undergoing the natural process of menopause." The FDA was even more dismayed that some women were taking estrogen "under the illusion that it will prolong their youth." Insisting "that women be informed and that they decide for themselves if the risks are worth the benefits," the FDA ruled that the packaging insert explicitly note that estrogen should be prescribed cyclically and that physicians should monitor patients closely. The insert should also say that after a patient had been on ERT for six months, the physician should either reduce the dose or halt the therapy. "The risk of cancer appears to be greater the longer estrogens are used and the higher the dose." Finally, estrogens should never be prescribed for enhancement purposes. "Estrogens should not be used to treat simple nervousness during menopause, because they have not been shown to be effective for that purpose. Neither have estrogens been shown to keep the skin soft or to keep women feeling young."[89]

The FDA response angered both pharmaceutical companies and several medical organizations, including the American College of Obstetrics and Gynecology.[90] Ayerst immediately dispatched a "Dear Doctor" letter to physicians all over the country stating that the *NEJM* studies did not establish a conclusive link between estrogen and cancer. The physicians, for their part, saw the FDA as interfering with the practice of medicine by spelling out what patients should be told about ERT. At a time when truth-telling about fatal illness was not the practice and informed consent got short shrift, doctors were not in the least embarrassed by an insistence that patients not receive relevant medical information. "This governmental intrusion into the physician-patient relationship," wrote one physician in *JAMA*, "is a serious matter indeed. . . . As in the case of virtually all pharmaceuticals, the medical profession and the pharmaceutical industry can be expected to continue to move toward more beneficial and relatively safer estrogen therapy programs." With no sense of irony he concluded that, in the search for better products, "The ethics of the profession and the profit motive provide the driving forces."[91]

Failing to dissuade the FDA, the Pharmaceutical Manufacturers Association and the American College of Obstetricians and Gynecologists took their case to Congress. When lobbying did not succeed, they went to federal court, charging that the FDA lacked the statutory authority to require patient package insert warnings. The federal court, however, upheld the FDA and the inserts remained.[92] The net effect of all these considerations was a sharp drop in the use of estrogen. Whereas, earlier, general practitioners had prescribed ERT as often as gynecologists, they now placed far fewer women on the regimen. And women, too, reduced their demand. In 1975, 27 million prescriptions for ERT were filled, in 1980, 14 million.[93]

V

In 1979, the NIH devoted one of its special consensus conferences, designed to resolve controversial questions in medical practice, to ERT. It was not a happy meeting. The consensus panel was dissatisfied with the number of clinical trials and the quality of the data. As best it could judge, ERT increased the chance of endometrial cancer by four to eight times, and it attributed a recent decline in the incidence of the cancer to the reduced use of ERT. On breast cancer the panel was agnostic; research was insufficient to draw a conclusion. As for benefits, it found that ERT did prevent the "dramatic loss of bone," but so did good diet, exercise, and calcium supplements. It was also persuaded that ERT significantly reduced the symptoms of menopause, but existing data did not justify its use in combating depression or preventing heart disease.

Faced with such mixed findings, the panel recommended, first, that each woman make her own calculus based on her degree of discomfort and tolerance for risk. (By 1979, because of the feminist attack on medicine as well as the prominence of death-and-dying issues following on the case of Karen Ann Quinlan, it was not so radical to propose trusting the patient.) The panel also recommended that women who wished to take ERT to reduce the symptoms of menopause use as little as possible for as short a time as possible. Estrogen should not be used for any other indication until more was known about its risks and benefits. Finally, the panel noted that preliminary evidence suggested that adding progesterone to estrogen might reduce the risk of cancer, but "the safety of this approach has not yet been established." Except for the women with severe symptoms "who can't make it through the menopause," ERT should be put on hold.[94]

Nevertheless, in the twenty years that followed the consensus conference, physicians pursued a far more aggressive strategy. Gynecologists changed the treatment regimen, the rationale for treatment, and even the name of the treatment. Estrogen replacement therapy became hormone replacement therapy (HRT), and the difference was the addition of progesterone. They now downplayed HRT's value as an anti-aging remedy and emphasized instead its unique contribution to preventive medicine. The changes did help to increase the number of prescriptions.[95] Over the 1990s, best estimates were that approximately one-third of postmenopausal women were taking hormonal therapy.[96]

Why did estrogen have nine lives? Physicians who advocated estrogen did not take the risk of cancer seriously. The risks to women using HRT were no different than the risks people took in daily life, such as driving on freeways. Moreover, risks were ostensibly offset by the prevention of cardiovascular disease. "Since the absolute risk of death from endometrial cancer," claimed one physician, "is relatively small while that from cardiovascular disease is relatively large, it may be that the net effect is a lower total mortality in estrogen users."[97] But proponents did not mention that the increased likelihood of endometrial cancer was demonstrated, but the likelihood of benefits against cardiac disease was not. All the while, doctors were hardly eager to inform patients that a drug they had prescribed increased the risk of cancer. Indeed, since endometrial cancer is a comparatively rare disease, the average physician might never have seen it among his ERT patients. Finally, since HRT did reduce the symptoms of menopause, it occupied a special niche in medical prescriptions. Hence, for all these reasons, gynecologists and endocrinologists were reluctant to abandon the regimen. HRT was the default option, innocent until proven guilty.

Researchers also had a stake in the purported benefits of HRT. In the 1960s and 1970s, the National Institutes of Health became the major funder of clinical investigation, both in terms of dollars and academic prestige. It so dominated the research enterprise that investigators had to work within its system. The NIH relied upon study groups composed of specialists to review grant applications. A proposal to investigate a compound or a disease would be closely reviewed by experts in pharmacology, epidemiology, and the disease itself. Peer review of grants did reduce the influence of politics or favoritism (remember, the NIH is funded by Congress), but it also shaped the pattern of scientific careers. Study groups demanded a greater mastery of smaller and smaller areas of study, giving the edge to the narrow expert, not the newcomer. They scored higher those investigators who had accumulated their data thoroughly and carefully, moving from one hypothesis to another. Thus, ambitious physician/investigators in endocrinology/gynecology would begin a career by investigating the effect of an estrogenic substance on bone formation, cardiovascular disease, or cognition. As their careers developed, they became more knowledgeable about the properties of the substance and its effects. Before long, they had a stake in emphasizing its potential benefits and minimizing its risks, both because a drug with limited action would not receive attention from study sections or from journal editors, and because they did not, for obvious reasons, want to devote their life's work to an also-ran drug. Moreover, NIH grants were short, two years to, at most, five. To obtain grant renewals depended on publishing results with positive findings, even if they were preliminary and suggestive. Indeed, were they positive, preliminary, and suggestive, a second grant, and later a third, would be forthcoming.

In this way, the NIH, perhaps inadvertently but surely effectively, encouraged investigators who came up with a negative finding to scrutinize their findings to see if there was one dose or one subgroup of patients that had yielded positive results. If not all the patients in the protocol had benefited, perhaps the younger or the older or the less sick or the more sick had. The one option investigators wished to avoid, at almost all costs, was to abandon the drug or the particular disease and have to begin the lengthy process of identifying a new compound, learning its properties and effects, and obtaining the preliminary results that would impress a study group. The investigator who started over might lose salary, laboratory space, and status. Deans and promotion committees in medical centers evaluated faculty by the number of NIH grants they received, known in the trade by the prefix to the grant number: RO1. Go without RO1s for a period, or not have enough RO1s in the first place, and promotion was out of the question. Thus, put bluntly but accurately: if estrogen was your drug, tinker with it, run more trials, find new uses, but do not give up on it.

The idea that HRT would keep women feminine forever did recede, replaced by a rationale of preventive medicine. "Estrogens," as Robert Greenblatt, an endocrinologist, argued, "will not smooth the wrinkled brow nor restore the sagging breast but may so improve the general health of women as to make the aging process gradual and more tolerable."[98] HRT would ward off osteoporosis and cardiac disease.

For osteoporosis, researchers compared users and non-users of estrogen and found that the regimen did inhibit bone loss and bone fracture.[99] These retrospective studies pointed to a 50 percent reduction in the risk of vertebral fracture, and a 25 to 30 percent reduction in hip fracture. The data were sufficiently compelling for the FDA, in 1986, to approve Premarin for the prevention of osteoporosis, with a recommendation that it be taken at a low dose, 0.625 milligram, for short periods of time.[100] No one had difficulty with a low dosage—the problem was with duration of treatment. Robert Lindsay, an endocrinologist who had studied estrogen and bone density for more than a decade, and most other bone and endocrine specialists, insisted that women had to be on the regimen for at least ten years. Because "on cessation of therapy, bone loss resumes," women would be at risk as soon as they went off the regimen.[101] Thus, women had to decide for themselves whether the risk of cancer was preferable to the possibility of spending their final years as a cripple.[102]

Osteoporosis was simple compared to HRT and cardiovascular disease. A protective effect, as investigators put it, was "biologically plausible," since premenopausal women suffered less cardiac disease than men, and postmenopausal women on the regimen had lower cholesterol levels than those who were not.[103] Moreover, observational studies, albeit not random clinical trials, had reported for decades that women taking ERT had less heart disease. The need for better data was apparent, and drug companies and the NIH, which in the 1990s was making women's health a priority, undertook the trials.

The first results were mixed. One month, a study confirmed effectiveness—women nurses who self-selected to take estrogen had less cardiac disease than women who did not. But the next month, another study contradicted it. The overall trend was distinctly negative. The HERS study (Heart and Estrogen/Progesterone Study), the first gold-standard, double-blind, placebo-based trial on HRT and cardiac disease, supported by Wyeth-Ayerst, reported in 1998 that 2,763 women with documented cardiac disease who took estrogen and progesterone fared no better than the placebo group. There were 172 coronary heart disease events in the HRT group and 176 in the placebo group.[104] Then the ERA (Estrogen Replacement and Atherosclerosis) study revealed that HRT did not stop the progression of hardening and blocking of the arteries in women taking HRT.[105]

What about HRT for women who did not have a history of coronary heart disease? Was it protective for them? Reviewing all the evidence, a summary article in the 2001 *NEJM* said no. "Given the absence of evidence from randomized trials that hormone-replacement therapy prevents heart disease, combined with the possibility of short-term harm, postmenopausal hormone-replacement therapy should not be prescribed for the express purpose of preventing coronary heart disease or cardiovascular events among healthy women, women with multiple coronary risk factors, or those with documented heart disease."[106] That same year, the American Heart Association issued its own, equally negative, recommendations: HRT should not be used to prevent recurrence of cardiac disease among women already suffering from it. And for those free of the disease, "there are insufficient data to suggest that HRT should be initiated for the sole purpose of primary prevention of CVD (cardiovascular disease)."[107]

VI

Despite the negative findings on the disease prevention quality of estrogen, the search for other possible benefits continued and even accelerated. The prime case in point was its potential to prevent memory loss and cognitive decline in older women. Although the underlying causes for these changes were unknown, epidemiological studies revealed that at every age level, women, more often than men, were diagnosed with dementia and Alzheimer's disease.[108] Some investigators hypothesized that the abrupt drop in estrogen production in postmenopausal women made them more vulnerable.[109] Perhaps estrogen supplementation would help to improve neuronal survival in the areas of the brain responsible for cognition, or to express the enzymes that are vital to memory, or to assist in clearing plaques from brain tissue.[110]

Animal studies lent some support. Rats who had their ovaries removed and then were given estrogen performed better than controls in learning to make their way through a maze. So too, rats in whom stroke was induced and who then received estrogen were better able to repair and minimize brain injury.[111] But the results in humans have been less impressive. A flurry of studies into the mid-1990s suggested the efficacy of estrogen against cognitive loss, that is, until one read the fine print. Some of the studies were retrospective, looking back to compare rates of dementia among women already using estrogen with those who were not. Users, it turned out, had lower rates of dementia than non-users, but here again, the two groups were not comparable: since estrogen users were better educated and healthier than others, these factors might account for the difference. Other studies emphasized that women using estrogen did better on one type of cognitive test, minimizing the fact that they do no better or perhaps worse on many other types. One study even gave estrogen high marks because women taking it scored higher on the second of two identical tests; the comparison group, however, was given the test only once, so the higher score may simply indicate that people score better as they become more familiar with a test. Still another study neglected to take into account the fact that its estrogen users were just entering menopause and were reporting marked relief from symptoms of menopause (hot flashes and insomnia), so their better cognitive scores probably reflected an improved physical condition and ability to concentrate.[112] When, in 1998, a team from the University of California at San Francisco (UCSF) reviewed all the prior research, its conclusion was that estrogen had no demonstrated preventive value against Alzheimer's.[113]

But researchers, as we have noted, were loathe to abandon estrogen. Despite its clear negative finding, the UCSF paper was actually cited three years later as "showing possible positive effects of HRT on over-all cognition," which occasioned yet another literature review.[114] The authors found it frustrating because different investigators used different methodologies, making it practically impossible to bring their findings together. Seventeen investigators, for example, used a total of forty tests to measure cognitive ability, but only seven were used by two or more teams, and thirty were used by one team each. Only two of the studies administered the same estrogen formula and dose. The papers together provided scant evidence for estrogen's cognitive effects among women who were not experiencing hot flashes, insomnia, and moodiness. To be sure, women who did have these symptoms and took estrogen scored higher than women with these symptoms who were not taking estrogen. It was not much of a finding—of course, people who are less moody and sleep better test better—but it was enough to sustain the authors' call, and the journal editor's call, for "continued intensive study."[115]

Is estrogen an effective antidote once Alzheimer's strikes? Again, a flurry of small studies, involving very few patients over short periods of time, initially suggested that it was. Preliminary findings inspired a full-scale investigation: 120 women with mild or moderate Alzheimer's were put into a twelve-month, placebo-control, blinded, and randomized study. They were divided into three groups: one received a placebo; the second, 0.625 milligram of estrogen (low dose); and the third, 1.25 milligrams of estrogen (high dose). The groups were then subjected to a battery of tests, including two types of tests for memory, three for attention, two for language and motor skills, and one for mood. The overall findings were negative: "The results of this study do not support the role of estrogen in the treatment of AD."[116]

Although one might expect the story of estrogen for Alzheimer's to end there, in fact, the research goes on, feeding off logic (since it should work, it will work), slim findings, persistent optimism, and new preparations. Thus, investigators respond to a negative finding by insisting that more data are needed (which is true of almost anything), or will suggest that even if estrogen cannot do much alone, it "could have an important role as an adjuvant treatment."[117] So too, as soon as a new estrogen-based product appears, like raloxifene, investigators rush to see whether it might improve memory or abstract reasoning or information processing. In fact, to answer the question about raloxifene, some 7,500 women at 178 sites in 25 countries received low-dose raloxifene or high-dose raloxifene or placebo, and were tested by six measures of cognitive function over three years. "The scores," reported the research team in the *NEJM*, "improved slightly in all three groups . . . with no significant differences among the groups."[118] How did the *NEJM* editorial treat the negative finding? First, it noted that the results were reassuring, for "there was no decline associated with raloxifene." Then it conceded that the results were disappointing since the groups receiving it did not perform better. It closed with a call for more research: "It would be premature to conclude that the use of estrogen-replacement therapy . . . has no role in preserving cognitive function."[119]

VII

How did women respond to these different findings? Variation by geography, age, social class, and race has been the rule. Use of HRT was higher in the South and the West, lower in the Northeast. Women aged fifty to fifty-nine were twice as likely to use hormonal therapy as those sixty to sixty-nine.[120] The better educated the woman, the more likely she was to be on HRT.[121] Women physicians were more likely than other women to go on HRT.[122] Women who identified themselves as black were 60 percent less likely to have ever used HRT than those who identified themselves as white.[123] Ambivalence among women about the regimen was also very high. Among all women who receive an HRT prescription, between 20 percent and 30 percent never filled it, and another 20 percent to 30 percent discontinued use within six to eight months.[124]

What about the reactions of professional medical societies? Through 2001, positions on HRT varied by specialty. The American Heart Association, as we have seen, was negative. The American College of Obstetricians and Gynecologists (ACOG) continued to favor its use and so did the American Association of Clinical Endocrinologists (AACE), with both minimizing risk and maximizing benefits. ACOG was convinced that although epidemiological studies suggested that HRT "may increase the risk of breast cancer, this increased risk has not been proved."[125] The AACE was also skeptical, insisting that two studies that associated HRT with breast cancer were misleading: "While there was a small increase in the incidence of breast cancers, both studies may exaggerate fear among women who may inappropriately abandon necessary therapy."[126] On the benefit side, ACOG, as of December 2001, "continues to recommend that HRT be considered as treatment to relieve vasomotor symptoms and genitourinary tract atrophy and to reduce the risk of osteoporosis and, potentially, cardiovascular disease."[127] AACE affirmed HRT's value for osteoporosis and for "ameliorating existing symptoms (hot flashes, vaginal dryness and emotional symptoms)." It also counseled physicians to explain to patients how HRT might be useful in retarding the aging process. When taking a clinical history, doctors should ask about the "frequency of intercourse, ease of arousal, libido, orgasm and dyspareunia." They should be certain to perform a "quality-of-life assessment" that includes "psychiatric history, pre-menopausal mood disorders, premenstrual dysphoria and cognitive functioning."[128] The answers might provide ample reason for prescribing HRT.

Then, in July 2002, the truly unexpected happened. The most sophisticated and extensive trial ever held to test the benefits and risks of hormone replacement therapies was stopped in midcourse because the risks seemed too great to allow it to continue. In 1993, the NIH had initiated a $628 million Women's Health Initiative (WHI), the centerpiece of which was HRT. Scheduled to be completed in 2005, the WHI trial planned to recruit 65,000 women to take hormone therapies, and their outcomes would be compared to 100,000 controls. These many years later, the question of the benefits and risks of using unopposed estrogen, estrogen-progestin, dietary modification (essentially a low-fat diet), and calcium and vitamin D supplementation would be resolved once and for all. We would learn whether HRT affected, for better or for worse, women's vulnerability to breast cancer, osteoporosis, and cardiovascular disease.[129]

The sponsors had been hopeful that the research would confirm the benefits of hormone therapy for women's health. But that was not the case. The study's data safety and monitoring board—which reviews ongoing results to make certain that nothing untoward is happening— halted the HRT arm of the research. "The risk–benefit profile found in this trial is not consistent with the requirements for a viable intervention for primary prevention of chronic diseases." Over one year, 10,000 women taking estrogen plus progestin compared with a placebo would experience 7 more coronary heart disease events, 8 more strokes, 8 more pulmonary blood clots, 8 more invasive breast cancers, 6 fewer colorectal cancers, and 5 fewer hip fractures. In effect, women taking estrogen plus progestin could expect 19 more serious negative events per year per 10,000 women than women taking a placebo.[130] The editorial in *JAMA* that followed on these findings unequivocally issued a negative recommendation: "The WHI provides an important health answer for generations of healthy postmenopausal women to come—do not use estrogen/progestin to prevent chronic disease."[131]

Press coverage was extensive. The story was on the front page of many newspapers and ran as the cover story in *Newsweek* and *Time*. Many physicians interviewed expressed shock. "This is a bombshell," one declared. "This is a dangerous drug," said another, adding that she planned to urge her patients to stop using it immediately. But others were more reserved, saying that they would reduce their reliance on the therapy and prescribe it mainly for short periods to relieve menopausal symptoms. And still others were angry that the monitoring board had halted the HRT arm of the trial. The decision gave too much weight to relative risk; since the negative outcomes were rare, the absolute risk for any one woman was low, less than 1 percent. And such was their commitment to estrogen that they still hoped other regimens or other products, such as natural estrogens or estrogen patches, would be safer and better. As always, more trials were necessary.

Reactions from medical societies ran true to form. The American Heart Association adopted an even tougher position, going from a 2001 recommendation that the data were too limited to support putting women on HRT for cardiac protection to a statement that women should neither start nor continue on HRT for cardiac protection.[132] On the opposite end of the spectrum, investigators and gynecological societies offered a variety of reasons to continue research and treatment. The WHI itself had swelled the number of estrogen researchers by funding forty study centers, each with a director and considerable staff. Not surprisingly, then, negative results inspired the design of new protocols. As Wulf Utian, himself an investigator as well as the president of the North American Menopause Society (like so many others heavily supported by pharmaceutical companies), unembarrassedly declared when he heard the news: "There are an awful lot of interests at stake here beyond women's health. There are investigators with research grants, NIH grants and grants from the pharmaceutical industry."[133] Utian also suggested that HRT might only promote the growth of cancers already present, not induce them in the first place. So too, the increased risk of heart attacks came only in the first year and a half of use, so that longtime users did not have to immediately stop the regimen. All of which led him to a rather startling conclusion: "It's not just a matter of what the data says. Truth is opinion."[134]

The American College of Obstetrics and Gynecology urged women to remain calm, assuring them that the risk to any one woman was very low. The North American Menopause Society noted that since the trial was stopped prematurely, it was unknown whether a longer trial would have found greater beneficial effects, including cognitive functioning and quality of life. The American Association of Clinical Endocrinologists insisted that each woman confer with her doctor "to determine whether or not HRT is right for her."[135] Some women, it contended, could receive significant benefit from the therapy, while others might have to worry about its potential risks. Statements of both organizations put the benefits before the risks, reminding women that the HRT regimen led to a 37 percent reduction in colorectal cancer rates and a 24 percent reduction in hip fracture rates.[136]

Ayerst, now Wyeth, immediately challenged the WHI finding. The day it appeared, the company, as it had in 1975, dispatched letters to 500,000 health care providers to remind them that the drug was very effective in relieving menopausal symptoms and carried fewer risks with short-term use. It also sent its detail men to visit "high prescribing" physicians to urge them not to drop the drug. A follow-up survey of a group of these physicians conducted by ImpactRx, a private research firm, revealed that the responses of gynecologists were, on the whole, favorable. Eighty-one percent believed that the media had sensationalized the findings. Almost half remained convinced that the benefits of HRT use still outweighed the risks, and another 26 percent believed risks and benefits to be about equal. Again, as in 1975, primary care physicians were more receptive to critiques of estrogen. Only 25 percent believed benefits outweighed risks, while another 26 percent believed the risks and benefits were equal.[137]

The bad news about HRT was unrelenting. In May 2003, researchers conducting the Women's Health Initiative Memory Study (WHIMS) on women 65 and older, reported that 4 years into the trial, 40 of the 2,229 women given the hormone, as opposed to 21 of the 2,303 given a placebo, were diagnosed with dementia. Hormone therapy doubled the risk.[138] Then, in June 2003, WHI investigators reported a 24 percent increase in breast cancer among women using estrogen plus progestin, an increase that appeared only in the third year of the trial.[139] An editorial accompanying the article found a "strong confirmation that the association between combined hormone therapy and breast cancer is causal," and, worse yet, the "ability of combined hormone therapy to decrease mammographic sensitivity creates an almost unique situation in which an agent increases the risk of developing a disease while simultaneously delaying its detection."[140]

Yet another article published in May 2003 cast aspersions on the long-held perception that taking hormone therapy made women feel better. Researchers administered quantitative quality-of-life tests to women when they entered the trial, again a year later, and, to a subset of the group, three years later. They found that the use of HRT did not have "a clinically meaningful effect on health-related quality of life."[141] By now, the press was attuned to such negative findings. As an editorial in the *New York Times* commented: Women who took HRT "did not feel any healthier or more vital than comparable women who took placebos, nor did they have more sexual pleasure. Compared with those in the placebo group, their minds were no clearer, their memories no better, and their mental health no different."[142]

But HRT has not been abandoned. The pharmaceutical companies have learned to respond to bad news. Wyeth had known in advance of publication that the data from the WHIMS study would be negative, and so even before "embargoed" copies of the article were distributed to the press, the company, as the *British Medical Journal* reported, "secretly briefed a number of medical societies about the results."[143] Wyeth wanted them to be prepared to field questions from the press and from patients. For example, they could note that all the women in the study were over sixty-five and suggest that hormonal therapy might be more effective in preventing cognitive decline in younger women. Or they could suggest that using patches or other dosages might produce better results.

Investigators have also kept HRT alive. So long as there are unanswered questions—whether about subsets of the population or new formulations of HRT—a research agenda remains and funds will be forthcoming for further studies. Perhaps, as one researcher mused, "genetic variations may render some women more susceptible to the harmful or beneficial effects of estrogen" and, given the current interest in pharmacogenetics, there is surely a hypothesis that warrants exploration.[144]

Thus, based on the historical record, it is likely that a growing number of physicians (drawn more from primary care than gynecology) will stop prescribing HRT. Drug company sales will markedly decline. (Stock analysts reported that prescriptions for the products had dropped 30 percent two days after the announcement.) But given the determination of the drug companies and the stakes of the investigators, sooner rather than later new studies will get under way and new products will come to market. Probably, a substantial number of women will not refill their prescriptions, particularly those who are not troubled by severe symptoms.[145] The open question that we will now turn to is whether they will take the lessons learned from estrogen into other areas of medical enhancement. Will they now use HRT as the model for guiding their use of plastic surgery and liposuction? Will they use the story of estrogen as a template for evaluating future genetic enhancements?

FIVE

The Body As Turf

NO ASPECT OF MEDICINE provides better insights into the promise and pitfalls of enhancement technologies than plastic surgery. It is enhancement at its most pure, with no suggestion, for example, that the procedure will improve cardiac functioning or protect bones. From the consumer perspective, women—who are the great majority of users—must decide what risks they are willing to incur in order to have a more shapely breast, an unlined face, or fewer inches of fat. But in making their choices, they typically focus on the benefits, ignoring risks almost completely. They demedicalize the procedures, whether in selecting a physician or calculating the odds. From the profession's perspective, plastic surgery demonstrates the extraordinary competition that a new enhancement technology sparks among different specialties. When no one group owns a procedure and when the market for it is lucrative, a variety of specialties try to capture a greater share. In the process, they not only publicize the procedure but also experiment with more powerful and riskier techniques. In the end, patient safety is compromised, although the patient may be the last one to know.

I

In initial appearance, plastic surgery was reconstructive surgery.[1] Following World War I, surgeons applied their skills and techniques to repair faces mutilated in battle or ravaged by diseases like syphilis. Severely wounded men, as one surgeon reported, "will undergo untold hardships to be restored to the normal."[2] Soon surgeons enlarged their domain. As the historian Sander Gilman observes, the growth of the specialty of plastic surgery coincided with the spread of a race science and racial laws that linked physiology to character, temperament, and intelligence. People were what they appeared to be. Accordingly, a small group of surgeons, themselves often outsiders, altered the physical marks of so-called racial inferiority, such as a Semitic nose. As Gilman explains, enabling people to look like everyone else became a sufficient justification for the procedure.[3]

Plastic surgeons next went on to modify the body for cosmetic purposes, to improve its shape and erase the signs of aging. "The defect may seem slight," one journalist explained in the 1930s, "but its psychological consequences may be serious enough to undermine the patient's mental health or economic security."[4] Feeling better and doing better expanded the specialty's reach, taking it from the severely maimed, diseased, and stigmatized to almost anyone. Indeed, by the 1970s, the indications for plastic surgery had become almost limitless, from drooping eyelids to imperfect noses to too full or too small breasts, to fat anywhere on the body.

II

In 1976, word began to spread among Parisian women that an obscure surgeon, Yves-Gerard Illouz, a solo practitioner in a working-class district, had developed a method for removing fatty deposits without leaving a scar. French surgeons who heard about it were disparaging, for the procedure violated established operating principles. But the women who wanted to get rid of unsightly deposits of fat were not. Within two years, over three hundred of them underwent the Illouz procedure and more were waiting their turn.

Illouz had been interested in "the problems of fat deposits and discordances of the body profile" for at least twenty years. He followed the surgical innovations and tested them on his patients.[5] In the 1960s, for example, Ivo Pitanguy, a renowned Brazilian plastic surgeon, used a traditional scalpel to make a large incision in the thighs and buttocks to excise fatty deposits.[6] Illouz tried it, but found that one deformity replaced another—a long red scar with bulges of fat on either side of the incision.[7] Giorgio Fischer, an American physician practicing in Rome, designed a blunt-bladed tool (a variation on the scalpel) to dig out fatty deposits. The tool chipped fat deposits, like a "mining machine in a quarry," which he then suctioned out. The technique, however, left a large cavity, a long scar, and sometimes caused hemorrhages and clots.[8]

Dissatisfied with these surgical approaches, Illouz turned to gynecological instruments, specifically the cannula used in performing suction abortions. He liked its flexible, hollow, and circular design and its vacuum aspirator.[9] Where the idea came from to use it for removing fat is unclear.[10] Perhaps, as some plastic surgeons have suggested, he was performing abortions himself and was familiar with it. For his part, Illouz insists that he wanted to please a young actress who kept asking him to remove a lipoma, a large deposit of fat, from her back. To satisfy her request and to not leave a scar, he had the inspiration of using a cannula and aspirating the fat.[11]

In 1978 and again in 1979, Illouz reported his results at the annual meetings of the French Society of Aesthetic Surgery. By using a cannula, he told his colleagues, he did not sever nerves or rupture blood vessels. The technique also allowed him to suction out fat without suctioning out blood and to see, by looking into the container of the vacuum aspirator, just how much fat he had removed. Moreover, the procedure did not require general anesthesia, making it suitable for office use.[12] But no one seemed impressed. His French colleagues were offended by his use of the tools of the abortionist. Worse yet, his technique of "closed dissection" violated surgery's commitment to "open dissection," and working in a visible field. They also complained that his methods were not scientific; he used a pinch test to locate the deposit and then marked the area with a crayon. Indeed, the entire technique troubled them. After making a small incision, Illouz guided the cannula, which contained a saline solution, through the layers of fat and created a series of tunnels or "honeycombs." This "blind subcutaneous sculpturing" earned him the label of "Dr. Mole."[13]

Professional disdain may have gone beyond a discomfort with

Illouz's methods to a discomfort with him. Born in Algeria to French-Jewish parents, Illouz went to France in the mid-1950s, began his studies at the Faculty of Medicine at Montpellier, and completed his medical degree and surgical qualifications in Paris.[14] Although he regularly performed plastic surgery, he was denied admission to the French society of plastic surgeons, which, like France itself, was hostile toward foreign-born Jews. "Between 1977 and 1982," Illouz remembered, "I worked alone, and few people gave much credit to my efforts."[15]

American plastic surgeons were among the first to appreciate Illouz's techniques. They did not share French xenophobia and were more open to innovation, particularly if it promised to be lucrative. A chance remark by a French friend prompted Norman Martin, a senior attending surgeon at Beverly Hills Medical Center, to interrupt his 1980 Paris vacation to meet Illouz. Observing his procedure was an epiphany for Martin; he was "witnessing a revolutionary new surgical technique." Because liposuction ran so counter to traditional surgical principles, Martin was unwilling to trust his initial reaction and returned to Illouz's clinic several times. At each visit, he marveled at the procedure, and at the number of women crowding into the consulting room. With Illouz's permission, he questioned the women and found a high level of satisfaction. Thinking that his colleagues might disregard a verbal report, Martin videotaped the procedure.[16]

Despite all Martin's efforts, American surgeons were wary. "The initial reaction of the North American community," he reported, "ranged from disinterest to skepticism." Convinced that they were wrong, Martin drew up a special patient consent form and obtained permission from the Beverly Hills Medical Center to perform the procedure. The women were far more enthusiastic than his colleagues. "Word spread quickly. Within weeks I was deluged with potential surgical candidates." Like Illouz, Martin only employed the technique to make an already well-contoured body more pleasing, eliminating specific and local fatty deposits in thin and well-proportioned women. Even so, between 1981 and 1984, he performed his procedure on 415 patients, each of whom had an average of five separate fatty deposits suctioned.[17]

Another convert was Gregory Hetter, a clinical professor of plastic surgery at the University of Nevada School of Medicine, with a large private practice in Las Vegas. Hetter heard about Illouz's technique from Martin and went to Paris in the winter of 1982 to observe it. The clinic, he noted, was near the Moulin Rouge, "a district well known to Americans who served their country in two wars."[18] But Hetter, too, was amazed by what he saw. Illouz's technique was "a leap to an entirely different medium . . . by far the most important advance in body contouring that had been made during my professional lifetime."[19]

Hetter also arranged for Illouz to describe his technique at the 1982 meeting of the American Society of Plastic and Reconstructive Surgeons. The event was standing room only and Illouz spoke to a highly receptive audience. "Dr. Illouz showed us a great variety of slides," one plastic surgeon noted, "most either overexposed or out of focus. It was obvious to all in the room, however, that we were witnessing the dawn of something completely different."[20] After Illouz's presentation, the Ad Hoc Committee on New Procedures of the American Society of Plastic and Reconstructive Surgeons (ASPRS) decided to go to Paris to evaluate the efficacy of suction-assisted fat removal. One member, Norman Hugo, chair of the Division of Plastic Surgery at the Columbia College of Physicians and Surgeons, was amazed at the ease of the procedure and the results that Illouz achieved, especially the absence of complications and the high level of patient demand and satisfaction.

In short order, the committee endorsed the Illouz procedure. "Suction lipectomy operations," its guidelines declared, "are not 'humbugs'; they are valid in select patients with select indications and with very stringent caveats."[21] The method was entirely creditable: "suction lipectomy by the Illouz blunt cannula method is a surgical procedure that is effective in trained and experienced hands and offers benefits which heretofore have been unavailable."[22] It effectively and efficiently removed fat from the knees, ankles, and arms. Postoperative results revealed no ruptured blood vessels or fat embolisms or damage to the veins in the legs or sloughing skin. The committee had one caveat: The procedure was not aesthetically perfect. There was often "some waviness of the skin surface because of an unequal removal of fat in the area of the operation."[23] Nevertheless, plastic surgeons now had "a relatively satisfactory surgical method for the treatment of fat outcroppings on otherwise slender patients."[24]

The committee emphasized the simplicity of the procedure. "The operation may be performed by any trained plastic surgeon who makes the effort. . . . This can be accomplished by a combination of observation, symposium, and clinical 'hands on' experience."[25] It also issued criteria for patient selection. The ideal candidate was young and slender. It did not rule out older and more obese women but recommended that the surgeon inform them that since their skin was "excessive and loose," there would be "contour irregularities . . . poor skin re-draping in return for better contouring in clothing." If the women accepted the trade-off, the surgeon could proceed.[26]

Recognizing the novelty of the procedure and the absence of data on long-term effects, the ASPRS committee proposed several questions for study. "Do fat cells that are mechanically removed ever recur, or indeed is the individual born with a set number of fat cells?" Are fat deposits genetically determined so that even diet and exercise cannot eliminate them? It was also interested in possible therapeutic benefits. Would fat removal improve circulation in the lower extremities and reduce the "wear and tear of the musculoskeletal system"?[27] The questions were the right ones, but as we shall see, neither investigators nor practitioners pursued them.

Illouz's experience suggested that the demand for liposuction might be greater than for other types of cosmetic surgery. From 1977 to 1981, he alone had performed 3,000 procedures.[28] Since it could be done in an office, the surgeon was freed from the constraints of operating room schedules. Moreover, most women had multiple deposits of fat removed in a single session, and surgeons could define each removal as a separate procedure. So if a woman had fat deposits removed from her thighs, buttocks, and abdomen in one session, she was billed for three procedures. With patients paying between $2,000 and $5,000 a surgical session, liposuction was a gold mine.[29]

Accordingly, the ASPRS worked to ensure that board-certified plastic surgeons monopolized the procedure. Given liposuction's simplicity, its departure from standard surgical techniques, and the likelihood of high consumer demand, it was to be expected that a variety of specialists would try to muscle in. Labeling potential competitors "quasi-practitioners or untrained physicians," the ASPRS predicted that they would "apply the procedure to minimal or inappropriate deformities for maximum reimbursement."[30] The only way to preserve the integrity of liposuction was to reserve it for plastic surgeons.

Monopolizing Illouz's technique was all the more important because the field of plastic surgery itself was becoming overcrowded. In 1943, there had been only 139 board-qualified plastic surgeons; by 1982, the number had increased to 2,500. In 1954, 31 hospitals had plastic surgery residency programs; by 1981, 106. In 1983 alone, the Board of Plastic Surgery certified 268 candidates, almost twice the number of plastic surgeons in practice only forty years earlier.[31] With so many qualified surgeons entering the field, how was the ASPRS to protect its members' incomes? Reconstructive surgery—the repair of birth deformities, serious injuries, and cancer excisions—did not supply enough cases to fill surgeons' schedules. Hence, they were turning to office plastic surgery, with estimates that it was already accounting for 80 percent of their income.[32] As two surgeons noted: "Their incentive was up-front cash for high-demand procedures, such as suction-assisted lipectomy."[33] If they did not capture a new and potentially lucrative procedure, the economic future of the specialty was in danger.

The ASPRS quickly organized liposuction courses and, in keeping with its ambitions, limited attendance to its members.[34] In 1984, board-certified plastic surgeons performed 56,000 procedures.[35] In 1986, the number was just short of 100,000, and by 1990, they were performing more than 125,000 procedures. Within six years, Gregory Hetter observed, liposuction "has risen from a belittled fringe procedure to the most commonly performed plastic surgery procedure."[36]

Despite their best efforts, board-certified plastic surgeons could not monopolize liposuction.[37] Illouz himself encouraged competition. His personal experience made him impatient with specialty control, and besides, he wanted to have the Illouz Method celebrated. He taught anyone and everyone how to do it—plastic surgeons, otolaryngologists, dermatologists, cosmetic surgeons, gynecologists, and general surgeons. In 1983, Illouz conducted seminars in six American cities as well as in Montreal, Rio de Janeiro, and Tokyo.[38] He also supported the establishment of the American Society of Lipo-Suction Surgery to teach and disseminate the technique to physicians, regardless of specialty training. Under its auspices, more than 200 physicians soon learned the Illouz Method.[39]

Competition also increased because, in 1975, the Federal Trade Commission (FTC) lifted the long-standing ban on physician advertising. Treating medicine as it would any other market enterprise, the FTC defined a prohibition on advertising as a restraint of trade and an effort to stifle competition.[40] The ASPRS and the AMA brought suit to reverse the decision, but the Supreme Court, in 1982, upheld the FTC.[41] The result was that physicians and surgeons were free to advertise their services, including their readiness to perform liposuction.[42] At first, some board-certified plastic surgeons refused to advertise. "Most of the advertising," one insisted, "is more suited to merchandising deodorants than surgical operations. The ads play on the frustrated vanities of people whose bodies may not measure up to the treasured ideals of *Playboy* centerfolds."[43] But as advertisements for plastic surgery began appearing in women's magazines, newspapers, and the Yellow Pages, almost everyone followed the trend.[44]

III

The most successful rival to plastic surgeons were dermatologists, and their entry into liposuction popularized the procedure. More competitors meant more publicity, more advertising, more patients, and not coincidentally, less attention to possible side effects. In the mid-1980s, dermatology was busily expanding its own domain. Practitioners not only diagnosed and treated a spectrum of skin diseases and lesions (including acne, psoriasis, and a variety of growths) but also used quasi-surgical techniques, such as curettage (scraping), electro-surgery (burning), and cryo-surgery (freezing) to excise tissue. Some dermatologists used local anesthesia in removing small sections of skin for biopsy purposes. They generally labeled these procedures "skin surgery," as though they were akin to general surgery. Soon, skin surgery was comprising one-third to one-half of dermatologists' practice.

Not surprisingly, dermatologists wanted to acquire still other surgical techniques to expand their practices and increase their incomes.

Theodore Tromovitch, a prominent dermatologic surgeon, considered the effort altogether legitimate. Patients "expect the dermatologist to treat their skin problem whether the treatment is medical or surgical."[45] But whatever the logic, gaining access to surgical training was difficult. Dermatology residents had to obtain special permission to rotate through surgical specialty clinics. Since each surgical subspecialty wanted first—and only?—to serve its own residents, dermatologists were at the bottom of the list. Most of them had to become autodidacts. "A fresh pig's foot tied to a piece of peg board," Tromovitch explained, "is a simple, yet instructive surgical model for practicing suturing techniques, routine excisions and flaps, even though porcine skin is thicker and less pliable than that of a human."[46] They could also improve their skin grafting skills by sedating a laboratory rat, tying it to a board, and sewing pieces of skin to different sites. Despite their exclusion from surgical theaters, or perhaps because of it, dermatologists developed innovative office-based procedures.

Determined to enter plastic surgery, dermatologists emphasized their special advantages. Trained to remove skin lesions, they could apply these skills to remove blemishes or any other unsightly marks in their offices and without using general anesthesia. "More patients have surgical procedures of some sort performed on them in a dermatologist's daily practice than in a surgeon's," boasted one dermatologic surgeon.[47] And with an office-based, anesthesia-free practice, the procedures were more convenient, quicker, less expensive, and, perhaps, even safer for patients.

Predictably, dermatologists defined liposuction as skin surgery and, therefore, part of their competence. By 1984, liposuction was included in dermatology residency training programs, and dermatology conventions included training courses on how to perform it. By 1986, dermatologists were confident that they were securing the territory. "In the hands of the dermatologic surgeon," one of them proudly noted, liposuction "has evolved so that large anatomic areas such as the abdomen and hips can be treated under local anesthesia in the physician's office with equal or greater safety and ease than under general anesthesia in the hospital."[48]

The competition between surgeons and dermatologists affected every aspect of liposuction: the proper setting for performing the procedure (hospital or office), the type of anesthesia (general or local), and the appropriateness of blood transfusions (necessary or not). Both specialties were so determined to own the procedure that they could not even agree on a name. Plastic surgeons called it *lipoplasty*—the suffix linked it to procedures that were part of the surgical armamentarium. Dermatologists chose *liposuction*, a term that took the procedure out of the hospital.

Plastic surgeons doggedly insisted that liposuction should be performed in a hospital operating room with a trained anesthesiologist present. They warned that removing large amounts of fat often required transfusions to replace lost blood and electrolytes. Dermatologists responded that plastic surgeons had "an overzealous apprehension of liposuction blood loss . . . and a resultant overuse of transfusion," which caused unnecessary postoperative complications.[49] Because dermatologists, by contrast, suctioned small amounts of fat over the course of several sessions, they limited blood loss and fluid shifts. Of course, dermatologists had no choice but to adopt these methods. "Only a few dermatologists," one of them observed in 1986, "have the capability of administering general anesthesia in their offices, and hospital politics being what they are, only a few dermatologists will be granted hospital privileges for doing this procedure under general anesthesia."[50] But dermatologists adeptly turned what seemed a disadvantage into an advantage.

A second and equally contentious issue dividing the two specialties was how much fat could liposuction safely remove. Both found the Illouz criteria for patient selection—slim and well-proportioned women—too restrictive and both wanted to offer the procedure to heavier women. The plastic surgeons thought that anesthesia and the availability of blood transfusions gave them an edge. Dermatologists, for their part, redesigned the size and shape of cannula and altered the composition of the local anesthesia so as to limit blood loss and suction out even more fat.

The technique that gave dermatologists their greatest advantage over plastic surgeons emerged in 1986, when Jeffrey Klein, a California dermatologist, devised the tumescent procedure to suction large amounts of fat from several areas in one session. He first infused the subcutaneous fat with an anesthetic to numb the area. "A large volume of very dilute epinephrine," Klein explained, "is infiltrated into a targeted fat compartment prior to lipo-suction, producing a swelling and firmness. This tumescence of fat permits an increased accuracy in lipo-suction and minimizes post-surgical irregularities or rippling of the skin." His method also minimized "blood loss, bruising and post operative soreness," and avoided the dents and waffling that frequently occurred with other methods.[51] Most important, as Klein often noted, "the anesthesia produced by the tumescent technique is so complete that it permits lipo-suction of large volumes of fat totally by local anesthesia, without IV sedation or narcotic analgesia."[52]

Although Klein's procedure departed from conventional practice in several ways and he had tested it on only twenty-six people, he insisted it was entirely safe. The risks of injecting large volumes of diluted lidocaine into subcutaneous fat gave him no pause, even though his technique violated standard guidelines on the amount of lidocaine that could be safely injected. The limit set by the manufacturer and approved by the FDA was 7 milligrams per kilogram of body weight. The maximum dose was never to exceed 500 milligrams. Klein, however, derisively referred to the guidelines as having "an aura of infallibility comparable to divine scripture."[53] In his twenty-six patients, he had used a mean lidocaine dose of 18.4 mg/kg, amounting to a total of 1250 milligrams, apparently with no ill effects. He insisted, incorrectly as it turned out, that when physicians suctioned out the fat, they were also suctioning out "a large fraction of the lidocaine dose."[54] After the procedure was completed, the remaining levels of lidocaine were ostensibly so low that they would not cause postoperative complications. When Klein initially described his new technique, some colleagues accused him of "medical malpractice."[55] The charge notwithstanding, he was soon increasing the lidocaine dose from 18.4 to 35 mg/kg of body weight. His contention always was that since no one had measured the absorption of lidocaine when it was injected directly into the subcutaneous fat, the established limits were irrelevant to the tumescent technique.[56]

Plastic surgeons first derided Klein, and then adopted his technique. "Dosages of lidocaine," Gerald Pitman, a prominent New York plastic surgeon, declared, "used during the tumescent technique are far in excess of doses recommended by the manufacturer. Yet, our patients and those of others have shown no significant toxic effects from lidocaine."[57] Other plastic surgeons concurred. "So, do I recommend the tumescent technique to accomplish liposuction contouring?" Scott Replogle asked rhetorically. "You bet."[58] By 1990, plastic surgeons routinely followed Klein, and Replogle explained why: "I use the tumescent technique because it works, seems to be safe, and seems to allow for better results, which are certainly more easily accomplished."[59]

Over the 1990s, both plastic surgeons and dermatologists increased the lidocaine load in their patients. In 1991, the American Academy of Dermatology declared that 35 mg/kg of body weight was safe,[60] but soon surgeons and dermatologists were using 55 mg/kg as the upper limit.[61] Anecdotal reports of safety with 70 to 90 mg/kg and even 110 mg/kg circulated. As the doses of lidocaine became higher, the amount of fat aspirated grew larger, and a new nomenclature took hold. There was "large-volume liposuction" (LVL), "megavolume liposuction" (MVL), and "gigantovolume liposuction" (GVL).[62] To appreciate the amount of fat aspirated, think of a bottle of wine that contains 750 milliliters. LVL was the equivalent of almost seven bottles; MVL, the equivalent of ten and a half bottles; and GVL, the equivalent of sixteen bottles.

An occasional journal editorial or letter to the editor cautioned that such large doses could be toxic, even fatal. Sooner or later the physician would run afoul of "the immovable science of drug disposition and the immutable laws of dose-effect drug toxicity. Pushing the envelope may be a heart-pounding thrill for a test pilot, but it's his life at risk—not that of the passengers."[63] Determining a truly safe level required well-controlled trials, but none were conducted. Specialty societies did not press the issue. After its declaration, the American Academy of Dermatology said nothing more about lidocaine levels for another ten years. The Plastic Surgery Society insisted that dosage level was less important than the training of the practitioner. "In the hands of a properly trained surgeon, suction-assisted lipectomy is normally a safe and effective means of surgically contouring localized fat deposits that do not respond to diet and exercise."

IV

In their ongoing competition with dermatologists, plastic surgeons tried to identify new ways to control the liposuction market. They took advantage of their access to hospital operating rooms to give patients a financial bargain without reducing their own fees. They combined liposuction with traditional plastic surgery, offering a combination abdominoplasty (tummy tuck) and liposuction. Selling them together became so common that it earned the sobriquet "marriage abdomino-plasty."[64] Plastic surgeons informed prospective patients that they could save them money because third-party payers often reimbursed the costs of abdominoplasty (provided the surgeon declared that the girth was harmful to the patient's health) and covered the cost of anesthesia, the operating room, and the hospital stay. The patient would then have to pay only the surgeon's liposuction fee. Plastic surgeons also recognized that liposuction was more conveniently and cheaply performed as an office-based procedure and so they followed dermatologists into the community. By 1995, almost half of the members of the ASPRS were reporting that they performed plastic surgery in an office facility.[65]

The offices themselves ran the gamut. Some surgeons rented a suite and might or might not hire a nurse. Others joined together to rent or buy an outpatient surgical center, sharing equipment and staff. Still others opened a medical hotel, which provided basic nursing, catering services, and overnight accommodations.[66] As the market became more competitive, surgeons spent more money decorating and furnishing the facilities. Some turned their suites into spa-like settings where consumers could also purchase other beauty services, including massages and facials. Still others made liposuction part of a fitness and anti-aging program.

Surgeons, unlike dermatologists who were already in the community, now had to learn how to organize and administer these outpatient facilities. They turned to marketing consultants who told them that they were in the beauty business, and so were obliged to hire an attractive and courteous staff, and give them complimentary plastic surgery. "Your staff," one consultant explained, "is an extension of your image."[67] Their well-contoured bodies were advertisements for the surgeon's skills and provided would-be patients with a firsthand testimonial of the benefits they could expect.

Plastic surgery journals also devoted attention to the business aspects of the practice. "Marketing," as one article explained, "is a process of learning who your customers (patients) are, what they need . . . and using that information to implement new . . . services."[68] Less likely than dermatologists to have a roster of loyal, returning patients, the surgeons hired consultants to develop a marketing strategy, at a cost of $10,000 to $25,000 a year. They learned how to organize and update a mailing list of all women who had ever consulted them and the importance of periodic newsletters describing new procedures. The consultants also urged surgeons to contact their patients post-surgery to get their feedback and make certain they were satisfied. "A happy patient, on average, will tell 3 or 4 other prospective patients about the experience, but an unhappy patient will spread the bad news to 20 or more patients."[69] To ensure a good practice, interpersonal relationships and mailing lists were as important, perhaps even more important, than the surgeon's technical skill.

Plastic surgeons hired publicity agents to contact the editors of the style sections in newspapers, local TV talk show hosts, and owners of luxury hair salons. They persuaded the media to visit the surgeon's office, learn about the procedures, and give him airtime. Agents also arranged to have fashion writers undergo liposuction at no charge and then do a story on the experience. The most coveted prize came each spring, when television producers typically aired a program reminding women that bikini season was near. The producers would select a surgeon (here the right agent was key) to perform liposuction on camera. The reward was his name running across the screen.

New business tactics emerged, like offering a wary woman a "come-on," that is, "do a quick one and then the patient is back for more."[70] Were she skeptical about liposuction, the surgeon might suggest "auxiliary liposuction," doing a small area and then adding some finishing touches without charge at a follow-up visit. Surgeons learned, too, to fit their schedules to their patients. "Lunchtime Cosmetic Surgery" allowed the consumer to schedule liposuction on the face or neck and go back to work. One surgeon arranged to have a group of his patients tell a television audience about their "lunchtime make-overs," and show off their "new look."[71]

Plastic surgeons soon realized that the demand for liposuction might cross socioeconomic lines if they could find ways to make the procedure affordable without violating the hallowed rule of the trade: always get payment in advance. As Robert Goldwyn, a prominent Boston plastic surgeon, warned his colleagues, because plastic surgery might disappoint and cause unhappiness, "we have learned the wisdom of pre-payment. . . . Perhaps because they have usually paid a considerable amount, they may feel the pressure to convince themselves and others that they are happy with what they have done."[72] One tactic was to have the patient finance the cost of the procedure with a low-interest loan. Indeed, by the mid-1990s, financial service companies had recognized that serving the $13.2 billion plastic surgery market could boost profits. So the surgeon's office staff informed prospective patients about the possibilities of financing and the cost of monthly payments.[73] They then filled out the application in the doctor's office, the office manager faxed the forms to the company, and five minutes later, most of them had been approved for a quick low-interest two-year loan. The surgeon, in return, gave the company a percentage of his surgical fee.[74]

The finance company also acted as a collection agency for the surgeon, to the point of harassing delinquent customers. A few plastic surgeons were wary of the practice. "We are told," Tolbert Wilkinson, a San Antonio surgeon, noted, "that we can disassociate ourselves from the collection agency but that just isn't true. Anyone who tried to collect bills for you on a routine basis is acting as your agent and in anyone's mind is representing you and proceeding with your approval. That's bad public relations and a bad medicolegal approach to . . . plastic surgery."[75] But business was business and the advice often went unheeded.

Some finance companies insisted that plastic surgeons share the risk if potential customers had a weak credit history. They would reimburse the surgeon for only 60 percent of his total fee; how much more he received depended on how faithfully the patient repaid the loan.[76] Many surgeons accepted the terms. "My theory," a Long Island surgeon told a reporter, "is that if I didn't offer financing, the patient would be in somebody else's office."[77]

A few practitioners worried that these changes were corrupting the specialty. Plastic surgery, Mark Gorney declared, "is being viewed as the last refuge of the free-booter or pirate who can charge whatever the traffic will bear."[78] Others were concerned that this intense reliance on marketing strategies encouraged deception. "Have some truth in advertising," Carl Manstein urged his colleagues. "Get rid of some of those pictures of the high-paid models who are surreptitiously disguised as postoperative patients. Include in the advertisements the surgeon's complication rate, the number of lawsuits, and surgical fees."[79] Some critics worried that surgeons might perform unnecessary or very risky procedures in order to cover their marketing costs. Manstein, for example, feared that financial considerations would "creep into the decision-making process of the surgeon. He or she needs to increase revenues to meet these increasing expenses, and I believe it clouds surgical judgment."[80]

All these considerations came together to make liposuction in outpatient settings the "wild, wild West of health care." There were no colleagues or oversight committees or government regulations to discipline an excessively aggressive surgeon. As one attorney described it: "When neither the private physician nor the private facility is required to meet regulations that would otherwise apply to surgery or anesthesia services provided in a hospital or accredited ambulatory surgical center, the marketplace rule of caveat emptor applies."[81] In fact, any physician, board-certified or not, could perform office-based surgery. "Physicians who are fully trained, marginally trained, or not trained at all," one plastic surgeon noted, "can do a procedure in the office without interference. The surgeon merely has to feel competent to do the procedure."[82] Under these circumstances, variation in practice was the rule. Some surgeons were so eager to perform a large number of procedures that they "cut corners" by not taking a medical history or conducting a physical examination or even ordering blood tests. Nor were they always careful to evaluate drug interactions. They would administer a large dose of lidocaine without finding out if the patient was taking antidepressants, oral contraceptives, or diet drugs. These compounds mixed with lidocaine could cause a major adverse event.[83]

The consent process was also haphazard in office-based practice. Each surgeon wrote his own consent forms and the information that was included in one might be omitted in another. Some surgeons used standard surgical hospital consent forms, which provided little specific information on liposuction. Others devised special forms but did not include possible adverse events. No review board examined the forms for accuracy or completeness. Robert Goldwyn did publish his liposuction consent form, which included information about the possibility of pain, numbness, and bruising, but it made no mention of some of the specific risks of the tumescent technique, including the possibility of fat embolisms, excessive fluid loss, or lidocaine toxicity.[84] Gerald Pitman prided himself on the amount of information he gave prospective patients and the time he spent discussing their concerns, including, he claimed, such alternatives to surgery as dieting and exercise. Pitman distributed a several-page "information letter," which described the procedure and included the possibility of pain, swelling, and bruising, "contour irregularities," and other "Complications and Untoward Results." But Pitman's basic message was not to worry and he blamed the press for "publicizing and even sensationalizing disastrous consequences following liposuction." He had investigated "all serious complications and/or deaths associated with liposuction" and "found that almost all patients who had a disastrous consequence following liposuction had either been operated on by unqualified practitioners or had had additional simultaneous surgical procedures performed that were more likely to have caused the complication."[85] In good hands, all would go well.

V

But all did not go well. As liposuction became an unsupervised, office-based procedure, the risk of death became a stark reality. The exact number of fatalities was unknown because reporting systems were crude and unreliable. But slowly, really very slowly, information on deaths by liposuction emerged. A 1989 congressional hearing, inspired by plastic surgeons trying to reduce competition from dermatologists, uncovered a total of eleven deaths in office-based plastic surgery. But no changes resulted. Plastic surgeons continued to insist that liposuction was entirely safe with a very low death rate of only 0.02 per 100,000 procedures. And this rate was apparently constant despite the increase in the number of the procedures and the use of large-volume aspiration. So too, in 1994, dermatological researchers questioned 1,778 colleagues on complications and death. Only 66 practitioners responded, and they reported just minor complications and no deaths. Unfazed by a 3 percent response rate, the researchers concluded that tumescent liposuction was an "exceptionally safe method of liposuction."[86]

It took until the end of the 1990s for the complacency to shatter. Mark Gorney, a former president of the ASPRS and the medical director of a professional liability insurance carrier, reported the findings of a survey of plastic surgeons and insurers that he had conducted: between January 1997 and October 1998, there had been fifty to sixty deaths associated with routine liposuction procedures. His findings were supplemented by data on deaths after liposuction reported by individual states. The California Department of Health reported ten deaths from liposuction between January 1997 and June 1998. That same year Florida reported six deaths.[87] In May 1999, physicians from the New York City Poison Control Center and the Office of the New York City Chief Medical Examiner identified five deaths after liposuction, and their findings took on especial prominence by appearing in the *NEJM*. "Tumescent liposuction is not a trivial procedure," they declared. "Drug absorption and drug interactions, fluid management, prothrombogenic factors, and liposuction volume should be evaluated for this popular procedure. . . . Deaths due to cosmetic surgery should be a matter of serious concern."[88]

In 2000, Frederick Grazer, a plastic surgeon, conducted a more comprehensive survey of side effects and fatalities.[89] Using a variety of tactics, including a header for a questionnaire that read: "Your Reply (return envelope enclosed) Can Help Save a Life!," he achieved a response rate from plastic surgeons of over 50 percent. The findings were startling. "The rounded mortality rate for liposuction surgery in the late 1990s hovers near 1 in 5000 (20 per 100,000)." Adult herniorrhaphy, Grazer noted, which was also an elective outpatient surgical procedure, "carries a manyfold smaller mortality incidence of just 3 per 100,000."[90] Grazer noted that the vast majority of the deaths (77 percent) occurred in patients whose surgery had been performed in an office or an ambulatory center. Almost one-quarter were caused by a pulmonary embolism, which may have resulted from lidocaine toxicity. Many of the complications occurred during the first night after surgery, when patients were at home and being looked after by relatives and friends who were "ill-equipped to diagnose—let alone treat—hidden fluid shifts, tenuous circulatory status, subtle respiratory depression and wound contamination."[91]

The reports of death rates spurred state medical boards and state boards of health, already concerned about the absence of oversight and standards in office-based surgery, to reevaluate their guidelines. A few state legislatures even enacted regulations addressing some of the most deleterious practices. The Medical Board of California recommended that surgeons' aspirations of more than 5,000 cc of fat take place only in a "hospital type setting." The California legislature, in its Cosmetics and Outpatient Surgery Patient Protection Act (1999), mandated reporting whenever patients were transferred from outpatient settings to a hospital or whenever a death occurred. Each facility had to have a written transfer agreement with a hospital, or a surgeon on staff with hospital privileges, or a detailed emergency procedure.[92] In Florida, a state medical board task force restricted the amount of fat that could be suctioned in a single office procedure to 2,000 cc, and restricted the amount of time allowed for an office-based surgery to six hours. But when surgeons protested these measures, the board backed down. It doubled the amount of fat that could be suctioned to 4,000 cc and increased the office operating time to eight hours.[93]

New York first tried to sidestep the issue. In 2000, the legislature considered a bill requiring surgeons in office-based settings to report to the commissioner of health all adverse events, complications, transfers of patients to hospitals, and deaths.[94] When it failed to pass, the commissioner of health established the Committee on Quality Assurance in Office-based Surgery. The committee then recommended that surgeons have a documented plan for transferring patients to hospitals in the case of an adverse event, but left it to the surgeons to define and implement the plan. In November 2001, the New York State Supreme Court voided the guidelines.[95]

Critics occasionally chided colleagues for their unwillingness to self-police. "A normal temperature and a valid credit card," complained Mark Gorney, "are poor criteria for any elective aesthetic operation."[96] But most plastic surgeons and dermatologists preferred to blame the other specialty for bad outcomes. One committee of plastic surgeons insisted that the problem lay with non-board-certified practitioners. It deplored "the lack of peer review and legal safeguards that has permitted non-surgeons and sub-specialists (whose education is limited to a specific part of the body) to perform procedures beyond their recognized areas of training."[97] "Liposuction is generally safe," insisted the ASPRS, "provided that . . . the physician has basic (core) accredited surgical training with special training in body contouring."[98] By contrast, dermatologists insisted that plastic surgeons had more adverse events than dermatologists, as evidenced by the greater number of malpractice suits filed against them.[99] An American Academy of Dermatology Task Force did concede that conclusive data was missing for many aspects of liposuction, including the optimal doses of lidocaine and the amount of fat that could be safely removed in an office. But it, too, maintained that the procedure was safe provided the practitioner was qualified.[100]

This rote response received its most telling critique in the 2000 report of the Institute of Medicine (IOM) on medical error. Reviewing all medical interventions, the IOM focused on the institutional and structural determinants of outcomes. "The majority of medical errors do not result from individual recklessness," it declared, "but from basic flaws in the way the health care system is organized." Even the most qualified physicians would make errors if they worked in a faulty system. If two very different drugs were packaged in look-alike containers, sooner or later a patient would receive the wrong drug. If dosage requirements were communicated through scrawled notes in a chart, sooner or later a patient would receive an incorrect dose. To reduce medical error, the IOM recommended a nationwide reporting of adverse events so that the data for identifying and correcting system faults would be available.[101] Recognizing how ambitious the plan was, and how it ran counter to many traditional medical practices, the IOM looked first to correct hospital practices. What about outpatient facilities? The problems here, it confessed, were even more intractable. In the community, "the high premium placed on medical autonomy and perfection and a historical lack of inter-professional cooperation and effective communication" made oversight and regulation practically impossible.[102]

VI

None of these considerations deterred women from undergoing liposuction. For them, the procedure was a welcome alternative to the tedious process of chronic dieting, fasting, purging, binging, or long hours of daily exercise. Liposuction recontoured sagging breasts, drooping jowls, and flabby abdomens. ("I can wear a bikini again and look like I am thirty," one fifty-five-year-old woman declared.) Gone were thunder thighs, saddlebag hips, and pole legs, replaced by a "lean mean body."[103] Small wonder that in 2001 Americans underwent almost 400,000 liposuction procedures. Men, whose use of plastic surgery has been growing, accounted for 20 percent of them.[104]

Consumers did not perceive liposuction as risky surgery. Indeed, many did not think of it as surgery at all. It was not necessarily performed in a hospital or under general anesthesia. There were no large wounds that required time and special care to heal and no sizable scars to conceal. The recovery time (judged by return to work) was often days, not weeks. It could be performed as serial surgery, thighs in one session, arms in another, abdomen still another, thereby reducing the recovery period and extending the financial payments. Moreover, consumers felt in control of the process. Their vision of the ideal body set the parameters. They determined sites where the fat should be removed, and how much should be removed. Surgeons were their handmaidens.

Risk was so distant from consumers' minds that they paid little attention to it in picking a surgeon. In an earlier era, when undergoing plastic
surgery was often a well-kept secret, women tended to select their
surgeons on the basis of credentials (hospital affiliation and stature—
chairmen were highly desirable).[105] Indeed, surgeons typically made the
aesthetic decisions, some of which were so formulaic that today, when a
slide of a sixty-year-old woman who had her nose reshaped appears,
plastic surgeons will jocularly call out "Long Island 1965," or "the
Joseph Nose." But in the 1970s, the patients' rights movement, which
promoted choice, and the women's movement, which encouraged control of one's own body, helped to bring plastic surgery out of the closet.
The benefits of liposuction were frequently extolled in women's magazines, local and national newspapers, and television talk shows. But the
changes did not promote considerations of risk. The media focused on
success stories, frequently including before-and-after photographs. The
fact that this information was often influenced by publicity agents and
paid for by surgeons did not trouble consumers. They themselves told
reporters and talk show hosts about the pleasures that accrued from a
surgical makeover. Unlike women undergoing surgery for breast cancer, they did not weigh the risks and benefits of different techniques.
Convinced that liposuction was not dangerous, they perceived it as
more akin to a spa procedure than a surgical one.

Accordingly, consumers often took advice from salon and spa staff
who now served as list-keepers of plastic surgeons. Or they would contact a surgeon they had read about in a magazine, or watched operate
on television. Or they would enter "liposuction" into a search engine
and get dozens of pages of Web addresses.[106] The Web sites often
belonged to individual surgeons who provided their credentials and
office addresses and some answers to commonly asked questions. Consumers visited LipoSite, which represents surgeons from forty-one
states as well from Canada and the United Kingdom. Click on a state or
country and a list of practitioners appears. Click on a name, and the
physician's credentials and the types of liposuction he or she performs
appears. The site also features 200 sets of before-and-after photographs
and information on the pre- and post-surgery weight of the woman as
well as the amount of fat removed. Consumers could enlarge the pho-

tographs and search for evidence of a scar.[107] What is rarely found on any of these pages is the risks of the procedures.

Consumers are not always alert to physicians' marketing strategies. They may be impressed when a surgeon uses a measuring tape to determine the total body fat in their arms or abdomen and calls off a number to his nurse. Presumably the post-liposuction measurement will be much smaller. One woman was delighted when a surgeon, after measuring each region with his tape, looked intently at her chin (which he had not measured), and gave it a little pinch. No words were exchanged but the woman knew that the surgeon intuited that she was troubled by a fatty deposit in her chin and wanted it suctioned out.

This type of surgeon-shopping is not tantamount to making an informed choice. Consumers rely on surgeon-initiated advertising and marketing strategies, not on information about how many liposuctions a surgeon has performed or the number of adverse events among his or her patients.[108] Even when the promotional literature mentions safety, it minimizes the possibility of adverse events. To get information on them, as we have seen, is not easy. Departments of health and specialty societies do not keep a publicly available database on office-based surgery. And women have not demanded them. The advocacy activities that are so notable in breast cancer have no counterparts in plastic surgery. This is personal, not political.

VII

Why this should be so emerges from interviews with women who have undergone liposuction. They are, for the most part, delighted with the technology and do not place it within a medical context. They chose to use it—they were not told by a physician that they must have it. This is about shape and form, not illness. Thus, some women wanted to be transformed to please their partners; others sought to improve their economic opportunities. Many others wanted a "subtle" makeover, a change "just obvious to yourself."[109] This attitude, they claimed, distinguished them from those who undergo other surgery "to correct something abhorrent, to obliterate the signs of aging, to obliterate themselves." The opinions of family and friends were largely irrelevant, and if partners opposed the procedure, it did not matter. ("He thinks I'm a little crazy. Why would a healthy person want to undergo surgery?") Neither opposition nor indifference altered their decision.

For these consumers, the personal pleasure of looking younger or more sexual was crucial. The full-length mirror was the significant reference point.[110] "Every morning before I had liposuction," one explained, "I would look in the mirror and feel so sad and now I'm not sad anymore."[111] When women described their bodies after liposuction, their delight was evident. One thirty-seven-year-old woman (five feet five and 120 pounds) underwent liposuction to eliminate the fat deposits on her thighs so that she could have "gorgeous legs." After liposuction she reported that her jeans once again fit better. The change, she admitted, was barely noticeable. She still did not have "gorgeous legs," but liposuction brought her closer and she wanted to return for more. Another woman underwent liposuction of her hips and thighs because she was too embarrassed by her body shape to dance nude for her husband. Despite her husband's protests, she underwent the procedure and felt more confident and sensual, even if her husband did not notice a difference.[112]

What did the women know about the risks? As they recall it, the surgeon warned them about the possibility of aesthetic imperfections, such as skin waviness, dimpling, and a slight lack of symmetry. Some remembered signing a consent form that contained a little information on other risks. One woman, the wife of a doctor, asked specifically about side effects and was told "not to worry." Any complication "would be minor and easily corrected." She suspected he was wrong, but went ahead anyway. "I wanted the procedure and did not want to challenge him."

Not surprisingly, the women were unprepared to manage post-liposuction complications. A few reported that the surgeon's nurse had told them that they might have some bleeding, cramplike pain, or fluid loss the first night home, but particularly those who underwent the tumescent technique were shocked by the searing pain and discharge of so much bloody fluid that they were forced to change soaked sheets several times during the night. "I was oozing so much and was so numb," one woman who had 2,300 cc of fat suctioned reported, "that I had to stay in bed for forty-eight hours. I had to send my husband to buy me some Depend diapers to wear." A forty-five-year-old woman who decided, despite the admonitions of her physician/husband, to undergo liposuction arranged to have four large fatty deposits removed, and then on the day of the surgery added three more. She and her husband were unprepared for managing the postoperative complications. The next day, wearing dark glasses to hide her facial bruises, she compared having liposuction to having an abortion. In both instances, she had been determined to remove an unwanted substance from her body; in both instances, a cannula had suctioned it out, and after both procedures, she was left to manage the aftereffects. Another woman admitted that she was bruised and sore for a month after the surgery. Her stomach was "swollen, lumpy, and black and blue"; she had to sleep on her side and felt "like someone had beaten me up." Nevertheless, she was far more intent on discussing her next procedure. Now that her stomach and thighs were slim, she wanted to do her knees and ankles. Despite the postoperative distress, she had the highest regard for her surgeon. After hearing about her discomfort, he sent her flowers.

VIII

Beyond these personal accounts, a variety of medical and social interpretations try to explain the appeal of enhancement through liposuction. To some psychiatrists, the behavior is evidence of underlying pathology. Having fat suctioned out becomes a form of bulimia, or masochism. One psychiatrist insists that "the surgery might serve to punish the body through cutting or mutilating it. The plastic surgeon then would represent the powerful, magically omnipotent parent who could transform the bad child into an acceptable being much as the fairy godmother transformed Cinderella." But relief is temporary and satisfaction a transient phenomenon. Unaware of the psychological dynamics, sooner or later the women will undergo another round of liposuction.

Psychiatrists also diagnose liposuction users as suffering from the newly identified disease of "body dysmorphic disorder." Katharine Phillips insists that millions of Americans, most of them women, are afflicted with this disorder. Its symptoms include "a preoccupation with an imagined defect in physical appearance in a normal-appearing person. . . . This preoccupation can be persistent and pervasive lead- ing to . . . repeated visits to dermatologists and plastic surgeons in an attempt to correct the imagined defect." Treatment with antidepressants would help them "avoid unnecessary cosmetic surgery."[113] If Phillips and others have their way, consumers would use Prozac, not lipo- suction.

But critics from other disciplines reject the linkage of plastic surgery to individual pathology, and define its attraction in terms of powerful cultural imperatives about the body. Several interpretations celebrate liposuction as an expression of individual will and autonomy. "An ide- ology of personal consumption," notes one sociologist, "presents indi- viduals as free to do their own thing, to construct their own little worlds in the private sphere."[114] Another puts liposuction into the mainstream of contemporary medicine. "Transplant surgery, pacemakers, in vitro fertilization and plastic surgery have joined the more routine techniques of dieting and exercise, offering the individual increasingly dramatic possibilities for taking his or her body in hand."[115]

Scholars who identify themselves as feminine theorists are far less sanguine. Plastic surgery, they maintain, oppresses women because it reflects a misdirected effort to achieve an impossible ideal of beauty. Kathy Davis, who has analyzed the experiences of Dutch women, con- cedes that plastic surgery can be "a resource for empowerment for an individual woman."[116] But the more critical consideration is the nega- tive consequences of the objectification of the female body. "From the sexualization of the female body in advertising to the mass rape of women in wartime, women's bodies have been subjected to processes of exploitation, inferiorization, exclusion, control and violence." Plastic surgery is one more example of how women's bodies have "been regu- lated, colonized, mutilated or violated."[117]

Because any one individual cannot affect the structural and cultural system that engenders female subordination and, inevitably, makes women feel ugly and inferior, Susan Bordo, a professor of English and women's studies, recognizes that the semblance of choice from understanding the body as plastic can be empowering. "In place of God the watchmaker, we now have ourselves, the master sculptors of that plastic." But she goes on to say that the freedom is a false one, entrapping women in "the sadly continuing social realities of dominance and subordination."[118] Plastic surgery becomes no different than foot binding, purdah, or the veil.

By contrast, feminist film critics take a much more affirmative approach, not so much interested in confronting patriarchy as undermining it. Collapsing the "virtual" and the "real," they focus on how women are portrayed in films and how spectators then understand and act on the celluloid images of the female body. As Annette Kuhn writes in her influential book *The Power of the Image:* "A good deal of the groomed beauty of the women in the glamour portraits comes from the fact that they are 'made up,' in the immediate sense that cosmetics have been applied to their bodies in order to enhance their existing qualities. But they are also 'made-up' in the sense that the images, rather than the women, are put together, constructed, even fabricated."[119] But rather than reject makeup as a violation of a feminine aesthetic of appearance, they reject the narrow vision of a natural body (it is as much a fabrication as high fashion), and instead applaud the satisfaction, and thrill, that fashion can provide. "Condemning cosmetics and romance-confession magazines," Jane Gaines, another feminist film critic, insists, "means that we dismiss the women who buy them. Wearing high culture blinders, we are unable to appreciate the strength of the allure, the richness of the fantasy, and the quality of the compensation."[120]

Feminist film critics are ready to pursue these arguments to an extreme. They scoff, for example, at interpretations of the anorexic woman as weak and self-destructive. What appears to be pathological is really independent behavior. "Anorexia," Noelle Casky maintains, "is the cultivation of a specific image as an *image*—it is a purely artificial creation and that is why it is so admired. Will alone produces it and maintains it against considerable odds."[121] So too, they glorify female bodybuilders for "entering totally uncharted cultural territory." "Gender confusion," Gaines notes, can turn into "gender eradication."[122]

In sum, there is no consensus on the meaning of enhancing the body.

Indeed, the very diversity of opinion means that enhancement will have both proponents and detractors. One school will condemn it, another will celebrate it. Since there is so much allure for consumers and so much financial gain for physicians, there is little doubt that surgical enhancement, like medical enhancement, will go forward without significant attention to risks.

SIX

Borrowed Manhood

DISCOVERING THE SECRETS OF THE MALE BODY, like the fascination with the female body, occupied a prominent place on the research agenda for the new disciplines of physiology and endocrinology. Just as investigators in the first decades of the twentieth century shifted the female principle from the uterus to the ovaries, so they moved the male principle from the sperm released by the testes to the hormone released by the testes, that is, testosterone. The identification and eventual isolation and synthesis of testosterone altered clinical practice, providing a remedy for admittedly rare cases of delayed sexual development. From the very start, however, physicians had grander ambitions. Testosterone promised to preserve and restore manhood, reinvigorate physical and intellectual capacities, ward off old age, and reverse senescence. As with estrogen, physicians dispensed the hormone without distinguishing between cure and enhancement or concerning themselves with its potential risks.

These similarities notwithstanding, crucial differences distinguished the medicine practiced on men from the medicine practiced on women. Physicians' attitudes toward the male body and men's attitudes toward their own bodies differed from women's in crucial ways. The medical gaze never focused quite as intently or narrowly on male-specific organs and men were not so receptive to a message of Masculine Forever. Hormone replacement therapy, with testosterone as the counterpart to estrogen, did not become a standard male regimen. Why it might have assumed such a position, and why ultimately it did not, illuminate a fascinating chapter in both the history of gender and the history of enhancement.

|

Well before the nineteenth century, both folklore and medicine had explored the sources of maleness, seeking ways to promote strength, vitality, and potency. Ancient physicians administered potions of ground-up animal testes to men, in the belief that the testes, as the anatomical seat of masculinity, contained substances that promoted or restored vigor. Variations on this practice continued over the centuries, with the popularity of such brews and nostrums reflecting an intense demand for the products and a hope in their efficacy. After all, as every farmer knew, the testes affected energy and muscularity; to castrate a rooster produced a capon—fatter, softer, and less active. To castrate an aggressive farm animal (a horse, dog, or bull) rendered him more docile and manageable. Indeed, popular lore recognized that men castrated— whether by accident or on purpose (to maintain, for example, the high pitch of their singing voice or to render them suitable to guard the harem)—lost their manly characteristics. They gained weight, tired more easily, and were less assertive. The logic, then, was obvious: if a loss of testicular function rendered men weak, surely a gain in testicular function would render them strong.

Underlying these associations was the idea that the visible product of the testes, sperm and seminal fluid, represented the male principle. Of course, this product was essential to reproduction but it was also linked, wrongly, to other male attributes. It was commonly believed that the production of sperm was necessary for the development of male secondary sexual characteristics, including muscle mass and genital growth—otherwise why would castration inhibit their development? By the same token, the production of sperm was tied to sexual perfor- mance. Castration ostensibly made men impotent, which was precisely why some criminal codes invoked it as a penalty for sexual offenders.

These propositions encouraged the idea that because sperm was the source of male prowess, it, therefore, should not be squandered.

Through the nineteenth century, an entrenched medical as well as cultural concept of "spermatic economy" warned men to limit the frequency of their discharges. Excessive sexual intercourse or, worse yet, frequent masturbation deprived the male body of its vital substance and rendered the abuser weak, sickly, or even mad. This notion of an essential and finite male fluid sustained a vision of maleness as static and fixed. Semen assured masculinity, but the body's reservoir of the substance could not be expanded. It had to be hoarded because it could not be replaced.

The knowledge that the testes produced something more than sperm and seminal fluid, that this additional substance, discharged into the bloodstream, was critical to sustaining manhood, and that the attributes of maleness might be more plastic than fixed, came very slowly. Although unappreciated for at least fifty years, the 1848 experiments of the German physician and zoologist Arnold Berthold showed the way. Like later physiologists who had no qualms about vivisection or seemingly unnatural research, Berthold surgically removed the testes from six young male fowl. He left two alone, and as expected, they grew into capons. With two others, he removed and then reattached one of their testes to their intestinal tissue (which was rich in blood). Both of them grew into typical cocks, growing combs, crowing, fighting with rivals, and, in his words, showing "the customary attention to the hens." With the last two, Berthold removed one testis from each, exchanged them, and replanted them in the intestinal tissue. Both of these fowls became cocks. From these clever, even daring, experiments, Berthold correctly concluded that the testes released a substance into the bloodstream that produced and maintained male characteristics.[1] Although his work went unnoticed, he had, in fact, demonstrated the biological power of an internal secretion from a ductless gland.

Some forty years later, the research, if it may called that, of Charles Édouard Brown-Séquard did revolutionize the field. Brown-Séquard had impeccable credentials in French medical-scientific circles derived from his mapping of the sensory pathways of the spinal cord and identifying some of the causes of epilepsy. Not until very late in his career, in 1889 when he was seventy-two years old, did he publish his first paper in the emerging field of endocrinology. And what a paper it was.

To his colleagues' amazement, Brown-Séquard reported that he had been injecting himself with testicular extracts. For several years he had been suffering from muscular weakness, growing fatigue, sleeplessness, and constipation; to find relief, he had taken a solution composed of testicular blood, testicular extracts, and seminal fluids from dogs and guinea pigs. The results over a three-week period were spectacular. He was now able to work long hours in his laboratory and then write a demanding paper. His muscle strength, as measured on a dynamometer, increased dramatically, his urinary jet stream was 25 percent longer, and his chronic constipation had disappeared.

Although now treated in the history of medicine as something of a buffoon, Brown-Séquard actually brought an impressive logic to his self-experiment. The intervention, as he explained, was based on what were at the time considered unimpeachable facts. First, men castrated before adulthood were "characterized by their general debility and their lack of intellectual and physical activity." Second, masturbation and other forms of sexual excesses produced debility. Third, young men who refrained from sexual activity and conserved their seminal fluid had exceptional physical and mental strength. Putting these observations together, Brown-Séquard reasoned that "in the seminal fluid, as secreted by the testicles, a substance or several substances exist which, entering the blood by resorption, have a most essential use in giving strength to the nervous system and other parts." Thus, weakness, as in his case the weakness of old age, might reflect the "gradual diminishing action of the spermatic glands," and were this deficiency corrected, physical and mental capacities would increase. But how was one to revive the spermatic glands? Brown-Séquard offered a formula: a solution of blood from the testicular veins of animals mixed with semen and the juice extracted from the crushed testicles of dogs and pigs.

Although such strange concoctions were hardly novel, Brown-Séquard's formula inspired laboratory research and clinical applications. If the professor of experimental medicine at the Collège de France advocated such an approach and supported it with objective data (on muscle strength and urinary streams), it might well prove effective. Moreover, Brown-Séquard was breaking new ground (or, more accurately, reviving Berthold's neglected observation) in identifying the testes as a gland that discharged an internal fluid. He did not yet have the facts or language quite right; he continued to call the second substance "sperm." But he was joining what had been old men's tales with the young field of endocrinology.[2]

A number of physicians rushed to administer the substances to their patients, and some reported remarkable success. One Parisian physician injected spermatic fluid into "three old men," aged fifty-four, fifty-six, and sixty-eight, and claimed rejuvenating effects. In the United States, William Hammond, a former surgeon general, reported that the preparation reduced pain, restored potency, and improved cardiac functioning in his patients. Encouraged by these findings, Brown-Séquard became more confident that his own physical improvement had a physiological, not psychological, cause, and soon he was publishing new formulas. Cut bull testicles into four or five slices, mix with one liter of glycerine, store for twenty-four hours turning frequently, wash in boiling water, pass the liquid through a paper filter, and then sterilize at 104 degrees.[3]

Physiologists, for their part, began to investigate the biology of male sexual characteristics. They castrated fowls, rabbits, frogs, and pigs to study the aftereffects. Some of them repeated Berthold's work, confirming that the testicles discharged substances into the bloodstream that maintained secondary male characteristics. Others, noting that removal of testes led to an elongation of bone and body structure, began to explore the role of the testes in regulating growth and metabolism. The most original work focused on the physiology of the testes, seeking to pinpoint precisely which tissues secreted which substance. The key findings came from the highly imaginative animal research conducted by two French investigators, Pol Bouin and Paul Ancel. They hypothesized that the testes were made up of two distinct types of tissues that fulfilled two different functions. One type produced sperm, and the other a male substance secreted directly into the blood. The sperm-producing tissue was responsible for fertility (the primary male characteristic); the other was responsible for generating and maintaining secondary male sexual characteristics. To test the theory, Bouin and Ancel tied off the sperm ducts on experimental animals, rendering them sterile, and found that physical size, strength, and mating instincts were unchanged. They went further, removing one testis and excising the sperm-producing tissue in the other; they then examined the altered testis and found that the remaining tissue, the so-called interstitial tissue, or Leydig cells, had doubled in size, and was capable of maintaining the animal's secondary sex characteristics. They concluded, accurately, that fertility and masculinity owed nothing to each other. Sterility and maleness were perfectly compatible. One could be fully a man but incapable of reproduction.[4]

Other investigators were no less ingenious in confirming and extending these findings. Two English researchers, S. G. Shattock and C. G. Seligmann, experimenting on sheep and fowl, blocked the tubes that carried sperm from testicle to penis, and observed that the now sterile animals retained masculine characteristics. Clearly, the testes were producing and secreting some other substance into the bloodstream that sustained male characteristics and they, along with Bouin and Ancel, located the source of production in the testes' interstitial tissue.[5] In fact, these findings confirmed the common observation that men with two undescended testicles could not produce sperm but had typical male secondary characteristics.[6]

Thus, by 1910, the science of male physiology had successfully distinguished sperm from this other "male principle." But the two substances did not generate the same excitement among researchers. The male fluid, not the spermatic fluid, captured almost all the attention. Part of the reason was the challenge and intrinsic difficulty of isolating this newly identified fluid, which was discharged internally, not externally. Even more important, (male) investigators were far more interested in masculinity than fertility. They defined the male not so much by his ability to reproduce but by his manliness. Secondary characteristics trumped primary ones.

Why this should have been so reflects a profound scientific and cultural differentiation of male and female identities. Reproduction belonged to women, and for most of the twentieth century the study of infertility or, for that matter, contraception belonged to the gynecologist. The preeminent male attributes involved physical activity, performance, and virility, to which fertility was irrelevant. These distinctions to mind, it became the task of the biological sciences to understand the

Pacific & Atlantic

FIG. 54. The long and short of it!

The circus is in town. The greatest feature is the "World's tallest and shortest man." This chapter explains why such freaks exist

FIG. 55. The child at the left is a cretin; at the right is shown the same child after one year of thyroid treatment. What changes do you see?[1]

Examples of "circus freaks" and "cretinism" in a 1928 high school health textbook, used to spark students' interest in endocrinology.

(Reprinted from N. B. Foster, "Diseases of the Thyroid Gland," Nelson Loose-Leaf Living Medicine, *Vol. III, Thomas Nelson and Sons.)*

Gertrude Atherton *(front row, extreme left)*
with Samuel Goldwyn and his other "eminent authors."

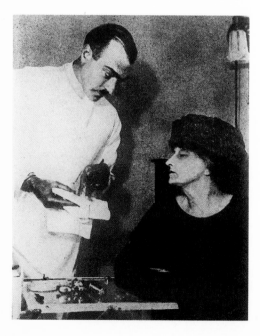

Gertrude Atherton consulting with
her own "gland doctor" after
Black Oxen made them famous.

(Copyright © Bettmann/CORBIS)

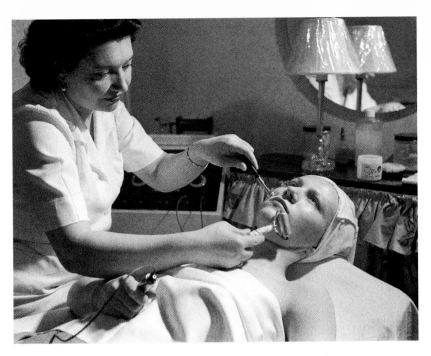

By 1944 facial rejuvenation was an accepted
surgical procedure and was carried out in hospitals.
(Copyright © Bettmann/CORBIS)

Drs. Harry Benjamin, Magnus Hirschfeld, and Max Thorak,
each committed to using medical skills to transform or rejuvenate the body.
(Courtesy of The Kinsey Institute for Research in Sex, Gender, and Reproduction)

H. G. Wells, the prolific and prescient
British author of science fiction.
(Copyright © Bettmann/CORBIS)

J.B.S. Haldane, one of England's most prominent and versatile scientists
and the author of *Daedalus, or Science and the Future*.
(Copyright © Mirrorpix.com)

Claude Bernard, a pioneer in endocrinology
and in the ethics of human experimentation.
(Copyright © Bettmann/CORBIS)

Eugen Steinach, who performed vasectomies on elderly men
in an effort to increase their supply of testosterone.
(Reprinted from George F. Corner, Rejuvenation, *New York, Thomas Seltzer, 1923.)*

Emil Novak, professor of gynecology at Johns Hopkins and author of one of the most widely used gynecology textbooks.
(Reprinted with permission from MedChi, the Maryland State Medical Society.)

Paul de Kruif, who helped popularize not only microbe hunters but also testosterone hunters.
(Courtesy of the Curtis Publishing Company.)

William Masters, one of the first and most influential proponents of hormonal therapy for postmenopausal women.
(Courtesy of Becker Medical Library, Washington University School of Medicine)

The Menopause: Its Modern Management

THE PATIENT approaching the menopause can now be reassured. Correction of endocrine imbalance relieves those distressing symptoms which may accompany the "change of life" period. This can be accomplished by administration of Theelin and Theelol, to compensate for the patient's diminishing secretion of ovarian follicular hormone. Successful treatment depends on careful adjustment of dosage to the individual patient's requirements. This in turn necessitates a preparation of precise potency and uniform purity.

PARKE-DAVIS THEELIN is a chemically pure crystalline substance. It is free from foreign protein and other extraneous matter. It is a naturally occurring estrogenic hormone, and is not chemically altered before administration. It is accurately standardized by both physiological and chemical methods. Parke-Davis Theelin (ketohydroxyestrin) is supplied for hypodermic administration in four potencies—200, 1000, 2000, and 10,000 international units per ampoule. The 1000, 2000, and 10,000 unit potencies are prepared in oil solution; the 200 unit potency, in aqueous solution. Theelin is also available in suppositories containing 200 international units.

PARKE-DAVIS THEELOL is a chemically pure, naturally occurring estrogenic hormone, accurately standardized by physiological and chemical methods. It is effective by mouth, and is useful in supporting Theelin treatment or, in mild cases, as the sole medication. Theelol (trihydroxyestrin) is available in Kapseals (hermetically sealed gelatin capsules) in two potencies, containing 0.06 mg. and 0.12 mg., respectively.

PARKE, DAVIS & COMPANY, DETROIT

The World's Largest Makers of Pharmaceutical and Biological Products

An early advertisement for estrogen by Parke, Davis & Company, 1937.

(Courtesy of Pfizer Consumer Healthcare, Pfizer, Inc.)

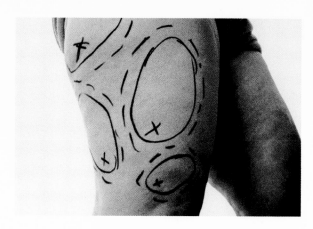

Preparing a woman for liposuction.
(Copyright © Bettmann/CORBIS)

The president of Genentech with company researchers,
just before Genentech began producing human growth hormone.
(Copyright © Roger Ressmeyer/CORBIS)

hormonal base of masculine capacities and the task of medicine to bring or restore capacities to young men of arrested development or old men of superannuated development.

II

No sooner was sperm tissue differentiated from interstitial tissue than an odd assortment of clinicians attempted to translate this finding into medical practice. One of the first was Eugen Steinach, who, like Brown-Séquard, was well trained in experimental physiology and well positioned within the profession. (Here, as elsewhere, credentials gave an unearned credibility to interventions—a dynamic that is still ongoing.) Born in 1861 in the Austrian Alps to a family steeped in medicine—both his father and grandfather were doctors—Steinach followed in their footsteps, taking his medical degree in 1886 from the University of Vienna. He then moved to the University of Prague and for the next twenty years conducted significant research on blood vessels and body muscles. With his reputation achieved (he became a full professor in 1907), he returned to Vienna and devoted his attention to hormones, particularly male hormones. His first investigations reviewed well-traveled territory. He demonstrated that when castrated rodents received a transplanted testicle, they reacquired their male characteristics; when they received a transplanted ovary, they assumed female characteristics. Along with Bouin and Ancel, he also located the source of male hormone in interstitial tissue.[7]

Had Steinach stopped there he would have secured a minor place in physiology textbooks. His fame, or his notoriety, came from what he did next. Steinach believed he had discovered the simplest and most effective clinical method for elevating the level of male hormone. Rather than follow Brown-Séquard's method of injecting the substance, he advocated the practice of vasoligation, not for the purpose of sterilization (what we now call vasectomy) but to increase male hormone production. "By a minor and absolutely harmless operation," he announced, "it is possible to activate, i.e., to stimulate to renewed vigorous endocrine activity, the patient's gonads."[8]

Steinach's logic was that by sealing off the vas deferens, the small tube through which sperm travel to the male organ, the amount of spermatic tissue would decline and, therefore, the amount of interstitial tissue would increase. It was as if the testes had an internal scale balancing the two functions; were the sperm side lowered, the hormone side would be elevated. The higher level of the male hormone would then "reactivate the entire endocrine system and organism."[9] By his own estimate, aging men would experience many benefits, the most important of which would be the prevention and cure of senility.

Steinach, like so many others, equated the symptoms of aging with the symptoms of castration. Forty- and fifty-year-old men first exhibited "presenile" symptoms, that is, "decreased libido and sexual potency." They next entered a senile state, experiencing fatigue, failing memory, indifference, and even depression. As physical debilitation and emotional and intellectual lethargy persisted, old men came to resemble old women, just as little boys and girls resembled each other. "The old man," wrote Steinach, "reveal[s] only traces of his former masculine aggressiveness and the old woman but feeble remnants of her former modest timidity and yielding gentleness."[10] Because every step in this degenerative process pointed to the effects of a decline in sex hormones, aging represented not a natural or inevitable state but a relatively easily diagnosed and curable illness.[11] Increase hormone levels and blood pressure and blood circulation would improve, as would strength, energy, and mental concentration, which Steinach believed was particularly important for the "intellectual class."[12] More, hormone therapy would remedy arteriosclerosis, adult-onset diabetes, depression, and prostate enlargement.[13]

Although these improvements came from a procedure that rendered the subject sterile, at least were it applied to both testicles, no one paid particular attention to this side effect. It seemed irrelevant insofar as the aged male was concerned, and almost irrelevant to the degree that younger men were more concerned about virility than fertility. And no one gave much thought to other possible risks. The idea that the decline in testosterone levels might protect the older male from diseases such as cancer had little currency. To the contrary. Physicians—it is difficult to know precisely how many—brought Steinach's findings right into the clinic. No matter that male hormone deficiency could not yet be accurately measured and no data confirmed that vasoligation actually raised levels of the male principle or of the other hormones. No matter that aging was not necessarily a disease or that interventions might well be enhancements. Doctors were facing unhappy patients, old men complaining of a reduced sex drive, impaired mental performance, fatigue, and malaise. Rather than just standing there, the doctors were doing something.

In the United States, no one more aggressively touted and employed Steinach's technique than Harry Benjamin. Remembered today as one of the first physicians to bring endocrinology to the treatment of transsexuals and persons undergoing sex-change operations—James/Jan Morris was one of his patients—Benjamin, who lived to be 101, was drawn to endocrinology because of its purported ability to treat aging. (He was, as we have noted, the very doctor who treated and inspired Gertrude Atherton.) Trained in Germany, he moved to the United States in 1913 at the age of twenty-eight and set up practice in New York City. In light of his German background and interest in hormones, he was among the first Americans to read Steinach's writings. Aware of the "boundless enthusiasm and exaggerated hopes" raised by the procedure and sensitive to the ethical question of whether Steinach was "interfering with the laws of Nature, and endangering our morals by producing roués," Benjamin traveled to Vienna in 1921 to meet him and review his cases. He came away totally convinced of the efficacy of the procedure—its outcomes were "striking and remarkable." In lectures and articles, mostly for medical audiences, he promoted the intervention, calling it not rejuvenation but the "surgical retardation of senility." "Vasoligature," he explained to New York colleagues, "causes atrophy of the spermatogenetic apparatus of the testes." Then, just as Steinach contended, the interstitial tissue expands, produces more of the male hormone, energizes the endocrine system and, in this way, prevents and postpones the symptoms of old age.[14]

Benjamin claimed that he had ample evidence to support the proposition. "I am satisfied to say," he declared, "that a regeneration and restoration of the vitality of younger years is actually possible for the aging organism." His first articles cited successful outcomes in Europe but soon he was presenting his own findings, reporting a success rate of more than 70 percent with seventy-five patients. Among them, he was delighted to note, "the intelligent class predominates by far," represented by doctors, authors, college professors, lawyers, musicians, and no fewer than twenty-three businessmen. Complaining that so many physicians adopted "an overcritical attitude towards everything new," he urged them instead "to follow some intuition in the practice of medicine in order to make headway."[15]

Presenile and senile patients, Benjamin emphasized, all shared the same symptoms, but they were generally overlooked by doctors precisely because they were so common.[16] To document the point, he published a series of case studies that included: "Physician, aged seventy-one. General decline. Frequent spells of dizziness. Loss of weight." "Salesman, aged fifty-five. Lack of pep and ambition. . . . Was frequently sleepy during the day. . . . Had to 'whip himself' to his work. Distinct decline in sexual ability." "Mechanic, aged forty-five. Drowsy and tired always." "Music teacher, aged sixty-nine. Felt distinct general and sexual decline. Wanted to restore his physical and mental strength." According to Benjamin, the Steinach procedure benefited all of them, producing weight gain, improved physical strength, increased sexual vigor, smoother skin, fewer wrinkles, reduced blood pressure, and, for the seventy-one-year-old physician, the ability to "study for longer periods with more satisfactory results." Benjamin insisted that side effects were nonexistent. The only possible risk was that the rejuvenated men would overexert themselves and suffer a stroke or cardiac failure. "The old man, who has regained some of his former vitality, is easily tempted to abuse his newly gained abilities and should most earnestly be warned not to change his mode of living for quite some time. He should enjoy his benefit moderately in work as well as in pleasure, and if he will heed this warning no harm can befall him."[17]

III

The benefits of elevating the levels of the male hormone seemed so substantial that physicians experimented with still other methods, including transplanting testes. Since testicular implants seemed to work so well with animals, they might be equally effective in humans. The first attempts involved desperate circumstances. In 1913, Victor Lespinasse, professor of surgery at the University of Chicago, reported a case of a thirty-three-year-old man who had lost both his testicles (one by accident, the other through botched hernia surgery) and was complaining of an inability to have sexual relations. Lespinasse decided to give him a testicle transplant, citing "the experimental work . . . done by Berthold." To his surprise, a testicle was "easily obtained"; although he gave no details, it probably involved a sale. The surgery went well, the testicle was implanted in the scrotum, and "the fourth day after the operation, the patient had a strong erection accompanied by a marked sexual desire." In fact, the patient "insisted on leaving the hospital to satisfy this desire."[18]

The largest series of testicular transplants occurred in California's San Quentin Prison under the direction of its chief physician, L. L. Stanley. Inspired by Brown-Séquard, he began in 1918 "engrafting human testicles from recently executed prisoners to senile recipients." After 1920, he switched to using animal testicles. Having performed 21 transplants with human testicles and over 300 with animal testicles, Stanley boasted of his excellent outcomes. San Quentin inmates were apparently lining up to get the procedure, and were subsequently reporting "a feeling of buoyancy, a joy of living, an increased energy . . . and mental activity."[19] Stanley went on to inject another 500 inmates with a substance made from animal testicles, and again reported positive results. The procedure alleviated such different complaints as nervousness, senility, sexual lassitude, and impotence, as well as acne, asthma, and rheumatism.[20]

Testicular transplantation also attracted publicity-hungry practitioners. Probably the most flamboyant was the Russian-born and French-educated Serge Voronoff. Following the path of Nobel Prize winner Alexis Carrel, Voronoff was drawn to experiments with organ transplants. When he was in his early fifties, he joined the faculty at the Collège de France, the institution of Bernard and Brown-Séquard, and combined his interest in transplantation with endocrinology. In 1919, Voronoff began transplanting monkey testicles into patients who complained of debility or loss of sexual drive. He announced his always favorable results in press interviews, not medical journals, and also wrote several popular books on the subject. The procedure, he declared, promoted not only physical and mental prowess but also had the potential to create a superior breed of human beings. Were bright young children grafted with testicles, they would grow up as a "new super-race of men of genius."[21]

The logic underlying testicular transplants and his own personal connections to Steinach and Voronoff encouraged Max Thorek, a surgeon at the American Hospital in Chicago, to establish "a little experimental station on the roof." A native of Hungary, Thorek was attracted, in his words, to the idea of "tampering with nature." After visiting the clinics of Steinach and Voronoff, he returned to Chicago to champion "therapeutic gonadal implantation" as a cure for "well-defined pathologic states."[22] The most important one was the male climacteric, manifested in "nervous, emotional, and vasomotor phenomena ... analogous to menopause symptoms."[23] Between 1919 and 1923, Thorek performed 97 testicular transplants; most of the testes came from apes and monkeys but some were from human cadavers. His results were good but by his own account inconsistent. The majority of his transplant patients (69) suffered from senility; after the procedure, 44 of them apparently returned to normal or were markedly improved, 25 were slightly improved or failures. Thorek also experimented with transplanting testicles into patients with mental illness, and about half of these patients, he believed, showed improvement.[24]

Thorek's readiness to tamper with nature led him to other interventions as well. He was among the first surgeons to do breast reduction and abdominal excisions (in the days before liposuction) to get rid of fat. His 1942 text on plastic surgery (one of the first to appear) justified these still very novel procedures. "In the business and social world," Thorek explained, "the individual afflicted with abnormal states of the breasts or abdominal wall is seriously handicapped," experiencing grave physical and psychological consequences no less serious than "congenital or acquired defects in other portions of the body. The need for scientific reconstructive surgery to remedy these handicaps is therefore evident." Like the endocrinologists, Thorek was convinced that raising the quotient of patient happiness was a legitimate medical task. "If surgery can restore happiness and enjoyment of life to an individual who has lost them, that is as strong a justification for its use as restoration to health."[25] The plastic surgeon who wrote the foreword to his text was even more intent on raising the quotient of happiness. The surgeon must remedy "the fixed fate" that nature has dispensed. "If the child can be given shapely ears he should have them for his own happiness; and who is to deny him that happiness if he can attain it?" If pendulous breasts restrict a woman's economic opportunities and "destroy her happiness, it is an obvious evasion of professional responsibility" not to treat her.[26]

Pharmacy joined surgery in the effort to realize the benefits of male hormones. Physicians prescribed an array of testicular extracts, including Henry Harrower's Gonad tablet. Not very imaginatively named, it contained 0.25 gram of adrenal, 0.50 gram of thyroid, 1 gram of pituitary, and 1.5 grams of prostate and Leydig cell extracts, and was to be taken three to eight times a day. Other preparations kept their precise formulas secret, but described them generally as mixtures of animal glands. The California Endocrine Foundation Laboratories, for example, sold Concentrated Orchitic Solution, which was one part tissue from the "small, hard testicular gland of the healthy young, live Goat, Ram, or Monkey," dissolved in a solution of alcohol and water. Testacoids, produced by Reed & Carnick, who also marketed Ovacoiods, contained testicular as well as prostate hormones.

How much standing did male hormone therapy have in its initial appearance? It all depended on who you asked. The intervention had just the right combination of major accomplishments, ambiguous findings, and outright fraud to animate all sides. Proponents cited the scientific credentials of the innovators and emphasized the logic of the procedure. If old men resembled castrates, then surely a hormone deficiency was the primary cause of aging, and a hormone supplementation was an apt remedy. Indeed, with medicine seeming to be on the cusp of an endless frontier, it was hardly far-fetched to imagine that it could reinvigorate the elderly. Others were distinctly uncomfortable with associating physical and mental prowess with a sexual gland and at least one physician cautioned that "testosterone must not be injected indiscriminately into every knave who aspires to emulate the sex behavior of the cock."[27] And still others wanted better data on whether hormonal deficiencies were real and whether hormonal supplementation actually worked. But these questions prompted still others: How much evidence of efficacy was evidence enough? Should a treatment's success with animals encourage physicians to try it on their patients? Or, put more broadly, when were laboratory findings properly translated into clinical practice, and who decided when to go forward?

These issues sparked an especially acrimonious debate between two camps, university-trained investigators and their specialist colleagues on the one side, and general physicians on the other. At stake, in the first instance, was the proper use and misuse of the new hormonal substances. When was it appropriate to be prescribing male hormones or other endocrine derivatives to patients? Under what conditions should the substances be given and with what degree of caution? Inseparable from these practical questions was the more general issue: Who set the standards for medical practice? Was it the clinician who drew upon his day-to-day experience or the university medical center researcher and specialist who drew on laboratory findings and clinical trials? How much credence should be given to the physician's report of the one patient who succeeded on a particular regimen, and how much to data scrupulously collected? In effect, decisions about treatment became enmeshed with debates over professional privileges and standing. Thus, the ordinary physician who prescribed hormones to his patients was either properly exercising the discretion of the general practitioner or not giving due deference to professional medical and scientific authorities.

The controversy permeated medical meetings, conferences, and journals. The most vigorous defense of the prerogatives of the general practitioner appeared in the pages of *The Endocrine Survey*. The journal was published by Dr. Henry Harrower, the founder of the California-based Harrower Laboratory Inc., which manufactured and sold (to physicians and druggists, but not to the public) the Gonad tablet. Harrower, who had spent two years in Europe studying hormones, had been among the founders of the first American medical society devoted to endocrinology and the first managing editor of its official publication, *Endocrinology*. But because of disagreements over hormonal therapies, Harrower broke with the society and began publishing *The Endocrine Survey*.

The case made in this journal for trusting to the ordinary practitioner rested on the value of "empiricism" within medicine, that is, the wisdom that physicians acquired in the day-to-day treatment of their patients. "Laboratory workers," complained *The Endocrine Survey*, "have extended their sphere of influence and have claimed the right to investigate and to develop methods of treatment that are taboo for the mere practitioner until permission has graciously been given."[28] If a therapy "has not been sanctioned in the laboratory," it is dismissed as empirical.[29] But "to disparage a new method of treatment on the plea that authorities have not sanctioned it is childish."[30] Ordinary practitioners were right to trust to their clinical experience and "employ a method of treatment that is not fully and definitively established"; in this way, patients would be benefited and knowledge about the intervention's strengths and weaknesses advanced.[31] After all, Jenner's vaccinations and Pasteur's vaccines were empirically established. Moreover, if physicians had waited until all the animal studies on the efficacy of thyroid extract against cretinism had been completed, they would not, even now, be prescribing this highly effective treatment.[32] "The only way to find out certain things is to—find them out."[33] Caution was its own vice. "If we wait in all doubtful cases . . . no new methods of fighting disease will ever be produced."[34]

Although the empirical school did not disparage all clinical trials, it objected strenuously to a system in which "treatment becomes simply a matter of laws, and the 'healing art' gives way to a healing science."[35] The ordinary clinician knew things that the clinical investigator did not, including the physical and psychological differences among patients. "No two persons react in exactly the same way . . . to the same disease of the same severity."[36] Never discount the "immense amount of information stored away in the head of . . . the mere doctor." So too, never discount what can be learned from a physician's report of a single patient who "suddenly benefits from a certain mode of treatment."[37] Bedside wisdom, the anecdote of a patient cured, was every bit as reliable as knowledge acquired through formal investigations.

The empiricists also attacked medical specialists who "limit themselves so absolutely to their own specialties as to forget that no patient
is merely a subject for the knife or exclusively a pair of sick eyes."
Specialists forget that "invariably the entire organism is sympatheti-
cally interfered with in its functioning." Hence, a "visit to the specialist
may possibly result in improved vision . . . but as long as the general
condition of the patient is not given attention, his health will not be
restored."[38] By contrast, the wise clinician recognized the value of studying the patient "from every possible angle. . . . The doctor cannot
afford, and does not have the right to aim at, only one organ in his therapeutic procedures." Specialists might complain about the general practitioner's resort to a variety of drugs, but prescribing only a single
preparation was to treat patients "more or less as experiment animals."
To the empiricist camp, laboratory physiologists were even less reliable
guides to proper clinical treatments. Investigators sought knowledge for
knowledge's sake, not for the relief of patients. They were "men who
have never treated a sick person and whose knowledge is confined
entirely to artificial conditions created in the laboratory."[39] By contrast,
"the aim of the practitioner of medicine is scientific . . . but only in so
far as the results . . . enable him to improve his means and methods of
relieving his patients of their illness."[40] With great pride, *The Endocrine
Survey* quoted Alexis Carrel's dictum that the task of the practicing
physician is "very much more difficult than that of the physiologist who
can select his problem . . . and solve it by . . . experiments." The good
physician was aided by science, but then "he has to guess. The great
clinician must possess the intuitive power of the man of genius."[41] In all,
"the proper study of mankind (the sick portion of mankind) is not the
guinea-pig or the rabbit or the white rat. It is man—sick man or woman
or child."[42]

If the manifest purpose of *The Endocrine Survey* was to elevate general physicians above specialists and researchers, the latent function was
to sell drugs. A publication underwritten by a drug company had an
obvious self-interest in encouraging physicians to dispense the new substances and to prescribe based on a single patient's success. Thus the
journal proclaimed: "We never cease witnessing new surprises concerning the remarkable and good effects that can be obtained from judicious
combinations of endocrine substances. The actual experiences of many

thousands of general practitioners have established this fact, and no university professor, no experimenter or research worker has the right to question these results."[43]

But question the results is exactly what researchers did, sometimes in anger, frequently with condescension. No practice distressed university investigators and specialists more than efforts by individual practitioners and drug houses to publicize and promote the uses of the male hormone. To Herbert Evans, the investigator who first demonstrated the link between the pituitary hormone and physical growth, endocrinology "suffered obstetric deformity in its very birth," because the midwives were the likes of Brown-Séquard and Steinach.[44] The American Medical Association's Council on Pharmacy and Chemistry complained bitterly about the "unwarranted and unsupported claims" that the companies issued for the male hormones; the preparations were "a menace" to sound medical practice. As another pioneering endocrinologist, Hans Lisser, insisted, a "drought descended upon the field," indeed, on the entire profession.[45] Hormonal therapies were "suspiciously sweet," noted the editors of the *Journal of the American Medical Association*. "How much longer will our profession continue to merit such criticism? Just so long as our profession continues to give serious criticism to pseudoscientific rubbish promulgated by the exploiters of organic extracts."[46]

Moreover, exaggerating the merits of hormone therapy undermined the use of therapeutic agents of "proven value." George Murray, who had demonstrated the efficacy of thyroid extracts in curing cretinism, decried how "little satisfactory evidence" supported the injection or ingestion of testicular substances. "To those of us who have devoted attention to endocrinology for many years, the recent exploitation of organotherapy for all kinds of diseases is deplorable, as it is apt to discredit a valuable means of treatment when properly employed."[47] The excesses also subverted sound research. As another *JAMA* editorial exclaimed, the ignorant physician "hinders medical progress by substituting simple faith and enthusiasm for the careful, critical study that is sorely needed."[48]

Despite their heated rhetoric, the defenders of research and specialization were actually in something of a bind. For one, they could not always agree among themselves on which therapies were truly effective.

For all the opprobrium heaped on Brown-Séquard and his successors, some physicians from reputable institutions, publishing in the major journals, believed that Leydig cells might proliferate after vasoligation, that ingested hormonal substances might galvanize the individual's endocrine system, and that testicular transplants might stimulate the patient's own glands to increase their output. Moreover, investigators had to be careful that in throwing out the snake oil, they did not inadvertently kill the snake—that is, endocrinology as a specialty had to survive the blind enthusiasm of general practitioners. Cast enough doubt on efficacy of hormone preparations and the ethics of their promoters, and the whole field might appear worthless.

Sensitive to such considerations, a number of specialists tried to stake out a middle ground. Lewellys Barker, a prominent endocrinologist at Johns Hopkins, was convinced that what passed for current therapy was unproven and haphazard. But he also appreciated how difficult it was for a general practitioner to resist patients' enthusiasm for a remedy that they had recently read or heard about, even if the supporting evidence was flimsy. Barker tried to set out good practice guidelines that would be acceptable to all sides, but his effort demonstrated how difficult, if not futile, it was to reach a compromise position.

Barker urged physicians to be more discriminating in their clinical decisions, to follow the "*principles,* or *laws,* that we believe to be well established, or that seem likely to become so." Physicians should attempt "to make complete diagnostic studies before they plan their therapy," and use "scientific imagination in devising better methods of application." In this way, they will be "doing their best to secure all that is obtainable for their patients, to help their colleagues in the profession by extending knowledge and improving technique, and to protect their fellows as well as themselves from any deserved opprobrium."[49] But even general practitioners who heeded such advice still had enormous discretion. They could freely dispense preparations that were "likely" to become, but had not yet become, "established." They were free to rely upon "scientific imagination," not just hard data, in making decisions. They were to think of themselves as part doctor and part researcher, which only opened the door wider to potential misuses.[50]

The physician-in-chief at Boston's prestigious Peter Bent Brigham

Hospital, Henry Christian, was trapped in the same bind. Writing in 1924, he confessed how easy it was to become either "over-enthusiastic" or excessively skeptical about endocrinology. "Somewhere between these two lies the happy mean." It was tempting to conclude that "this or that gland of internal secretion is hyperfunctioning or hypofunctioning or dysfunctioning when there is some superficial resemblance . . . to changes noted in animals when some gland of internal secretion has been in some way disturbed." But physicians needed specific evidence of a link between the pathological changes and a hormone deficiency and compelling evidence of the efficacy of a particular preparation. Most interventions, albeit not all, failed these two tests. Certainly in the case of male hormones, "past use justifies no confidence in any efficiency for testicular extract." At the moment, Christian argued, "the abuse of endocrinology is out-weighing the use among our practitioners." Nevertheless, the promise of marvelous discoveries—insulin was his case in point—made it imperative that "experiments should continue and new preparations be tested out. . . . Could the active principles of more of the glands be prepared in such form that they could be introduced into the body without losing their activity, we would have new therapeutic agents of very great value. . . . Because of so many failures, were we to become skeptical of any new claims and stop testing them, progress would inevitably end."[51]

IV

With the male as with the female principle, the laboratory breakthrough in isolating and synthesizing the hormone came over the decade 1925 to 1935. If anything, the dubious nature of the existing preparations made the search for the active substance more intense. Products like those distributed by the Harrower Laboratories did not drive out good research. Gresham's law did not operate in science.

If physiology first undertook the challenge of identifying the male hormone, biochemistry completed it. The field has its own special history. Between 1900 and 1920, it separated itself from general chemistry (which paid little attention to biological substances), medical chemistry (which concerned itself with toxins), and physiological chemistry (which was exclusively devoted to animals). By 1920, the leading medical schools all had departments of biochemistry, and within them, one of the highest priorities went to isolating and synthesizing the active agents in hormones.[52]

Although confronting formidable technical challenges, investigators made steady progress. In 1927, a biochemistry graduate student at the University of Chicago, Lemuel McGee, working under the direction of Fred Koch, managed to extract from freshly ground testicles of bulls a substance that when given to capons spurred the growth of male combs. Bringing an entirely new level of precision to hormone research, Koch together with his colleagues devised "test measures for detecting the presence or absence of the male hormone" by calculating its effects (in centimeters) on cocks' combs and photographing the different stages of growth.[53]

The tools of biochemistry and exacting standards made it apparent that none of the drug company male compounds had efficacy. "I am unaware," declared Koch in 1930, "of any active hormone preparation on the market." Biochemical analysis also put to rest any claims for efficacy of Brown-Séquard's recipes or Steinach's procedure. Brown-Séquard's preparations had been dissolved in water but it turned out that the active agent of the male hormone dissolved only in fat. So too, Leydig cells did not increase their production of the male hormone if sperm cells were eliminated. Nor was testicular transplantation any more useful. "We have as yet," concluded one biochemist, "no evidence that the introduction of foreign testis tissue has any effect." In all, reviewing forty years of preparations and methods, the conclusion was inescapable: "The clinical application of the testis hormone is highly questionable. . . . The idea of increasing the life span and usefulness has unfortunately crept into the general concept of the function of male hormone without any apparent basis of fact."[54]

But no sooner was the verdict issued than it was outdated. In 1929, several investigators discovered that male urine contained the male hormone. In 1931, Adolf Butenandt, on the faculty of the University of Göttingen and cooperating with the Schering pharmaceutical company, purified and identified minute amounts of the substance, naming it androsterone. (For this feat, he was co-winner of the 1939 Nobel Prize in chemistry.)[55] Then, in 1934, Leopold Ruzicka, a biochemist whose earlier work had involved synthesizing the chemicals essential to odors (which was of great value to the perfume industry), purified the hormone from cholesterol (and shared the 1938 Nobel Prize with Butenandt). Finally, in 1935, Ernst Laqueur and his colleagues at the University of Amsterdam, with the help of a Dutch pharmaceutical company, Organon, purified an even more powerful version of the hormone from bull testes and called it testosterone.[56]

V

The newfound ability to produce the male hormone in the laboratory, albeit not without considerable expense and difficulty, sparked an even more zealous effort to establish its clinical uses. Between 1936 and 1939, over two hundred articles on the use of testosterone appeared. Earlier negative findings now became irrelevant because testosterone, particularly when chemists reformulated it as testosterone propionate (which was some four times more powerful than the original), was a highly active agent. But active to what end? Would the compound be able to correct for sexual underdevelopment? Would it finally become possible to realize the ambitions of Brown-Séquard and his followers, and reverse the disabilities of old age?

The most unambiguous answer was that testosterone could restore masculinity to young men deprived of it because of a birth defect, accident, or illness. Like insulin, it could effectively compensate for a marked deficiency or total absence of the naturally produced hormone. One of the first and often repeated accounts of a successful intervention came in 1937 from James Hamilton, a physiologist at the Albany Medical College. Justifying his report of a case of one because testosterone was new on the market, Hamilton told of a twenty-seven-year-old male medical student who was engaged to be married. He complained of hot flashes, fatigue, sexual incapacity, and "social stigmata due to feminine aspect and high voice." In appearance, the student resembled a "pre-pubertal castrate," with wide hips, an absence of body hair, and underdeveloped genitals. In presentation, he was "intelligent, industrious, trustworthy, prone to anxiety, and feeling somewhat inferior because of his condition." Hamilton gave injections of testosterone, and within three days, he was reporting erections; within two weeks, his genitals grew larger, the hot flashes disappeared, his energy level increased, and he seemed "more self-assured and in higher spirits." Then, without informing him, Hamilton substituted a placebo for the testosterone, and the earlier symptoms returned. When Hamilton resumed the testosterone treatment, his symptoms disappeared. Confident that the changes were physiological and not psychological, Hamilton concluded that testosterone was an effective treatment for sexual underdevelopment.[57]

Over the next several years, physicians frequently reported success in such cases. "With the synthesis of testosterone," observed one researcher, "there ended one of the most amazing chapters in the history of biomedical research." And *amazing* was the term commonly invoked. Clinicians recounted how one patient underwent an "amazing change in his personality." Another patient became "jubilant."[58] The drug was equally effective with men who had been castrated. As soon as he could obtain testosterone (from the Schering drug company), Walter Kearns, a urologist and biochemist at Marquette University, contacted some of his former patients, "broken men," he called them, "nervous, apprehensive, depressed, unable to concentrate, devoid of libido." He gave them testosterone, which "produced a change . . . as definite and pleasing as anything in my experience." Five milligrams injected twice a week "increased strength and endurance, a desire to expand their work, the appearance of libido, erections, ejaculations, ability to copulate and an increase in growth of the beard.[59]

Case reports were often published with accompanying photographs, arranged as "before" and "after."[60] Some of them were body shots, demonstrating that once feminine shapes had given way to a more lean and muscular appearance. The hallmark of the photographs were close-ups of the genitals. The accompanying text gave the precise measurements of change, one centimeter before, four centimeters after. Whatever concern physicians might have had for patient privacy— many faces were blocked out but a sizable minority were not—the overarching need was to present incontrovertible evidence of physiological change after treatment. Having suffered through the days of Brown-Séquard and Steinach, investigators were insistent on demonstrating efficacy, and they did so by following the tradition of showing combs growing on cocks. Testosterone had restored manhood and confirmation was to be found in the pictures of enlarged genitals.[61]

The treatment did have limits. Investigators understood that testosterone could not restore fertility; in fact, it reduced the production of sperm. They realized, too, that it could not cure impotence or erectile dysfunction, except where it exerted a placebo effect. They also reported (incorrectly, as it turned out) that testosterone administered to individuals with normal levels of the hormone exerted no effects. And they believed, again wrongly, that testosterone had no side effects. "No undesirable actions have been observed in any cases treated with testosterone," noted the endocrinologist Henry Turner. This finding was particularly important because it meant that physicians could give patients a trial of the drug even without knowing whether they were or were not actually deficient in the hormone. The patient would start on testosterone and, if benefits followed, then apparently a deficiency existed and the treatment would be continued. If it did not bring benefits, it could be stopped, apparently with no adverse consequences.[62]

These successes acknowledged, the open question was whether testosterone was effective as an anti-aging compound. Inevitably, the successful treatment of young boys and men who lacked adequate levels of testosterone led to attempts to treat older men who might also have an analogous condition.[63] Since testosterone corrected a pathological state might it correct an undesirable state? Could testosterone not only cure but enhance?

No one was more certain of its ability to do that than Paul de Kruif. Famous for his book *Microbe Hunters* (1926), which probably inspired more students to pursue science and medicine than any other single volume, de Kruif in 1945 published something of a sequel, this one devoted to the hormone hunters. One hormone in particular, testosterone, was the focus of the book and supplied its title: *The Male Hormone*. The previous year de Kruif had written a brief article for *Reader's Digest* entitled "Can Man's Prime Be Prolonged?" His new book gave the answer:

an unqualified yes. The section headings outlined the argument: Growing Old Isn't Natural. The Rescue of Broken Men. Testosterone the Builder. A New Lease on Life.[64]

The physician whom de Kruif credited for inspiring his account was Herman Bundesen, president of the Chicago Board of Health. According to de Kruif, Bundesen was extraordinarily active at the age of sixty-two, mounting public health campaigns to eradicate one or another disease, writing papers, and generally pursuing life to its fullest. Although de Kruif was coy about it, there was little doubt that Bundesen was using testosterone, and proud of it. "The chemists suspect that growing old simply isn't natural," de Kruif quoted him as saying. "They're suspecting that growing old the way we do now is really only another disease." Where to find the cure? Bundesen knew the answer: "Don't sell these new organic chemists short."[65]

The case for testosterone seemed to de Kruif, as to others before him, irresistible. Once again, logic trumped the need for data. "If testosterone brought back muscular strength, mental keenness, joy in life to broken men without testicles, bringing them back to normal, wasn't it likely that this same testosterone did play a role in keeping up the muscle and brain power of normal men?" And since testicular activity apparently waned with age, "what was so cockeyed about supposing that injections of testosterone might replace what the aging testicles were no longer so active in producing?"[66] Because testosterone "could transform feeble and fuzzy-witted eunuchoid humans into alert and hard-working males," surely it could do the same for the elderly.[67] The reasoning seemed so self-evident that de Kruif attributed physicians' reluctance to dispense testosterone to personal and professional timidity. They were frightened to unleash "this new sexual T.N.T.," too worried about "what kind of spectacle would human oldsters make of themselves, with their banked sexual fires flaring again, with them prancing about under testosterone's hot influence?"[68] De Kruif himself had no such concerns and, at the age of fifty-four, was ready to trust himself to endocrinology in general and testosterone in particular. "Now I see a gleam of hope of science to help me extend the prime of my life. . . . So, no different than a good diabetic child who knows that insulin every day makes the difference between living and dying, I'll be faithful and remember to take my twenty or thirty milligrams a day of testosterone. . . . It's borrowed manhood. It's borrowed time. But, just the same, it's what makes bulls bulls."[69]

However chatty de Kruif's style or simplistic his argument, testosterone as an anti-aging compound had some support in the medical literature. For example, Allan Kenyon and colleagues from the University of Chicago School of Medicine, with funding from the National Research Council, demonstrated in 1942 that testosterone "exerts metabolic influences which are compatible with the formation of new nongenital tissue." Moreover, the new tissue was deposited in the muscle, thereby likely increasing body mass and strength in the elderly. In other words, Brown-Séquard was right in theory but not in practice. His "guiding idea that the testes might play some decisive role in somatic senescence has never been entirely lost sight of."[70]

Although the publicity given to de Kruif's claims irritated a number of investigators, their own work actually confirmed some of his propositions. Carl Heller and Gordon Myers, publishing in *JAMA,* complained that "physicians are deluged with requests for treatment by hopeful readers." But their findings supported the efficacy of testosterone in a subcategory of elderly men. Heller and Myers devised a test for measuring testosterone levels (using a combination of urine analysis and testicular biopsy) to help determine whether male climacteric or male menopause was a valid diagnostic entity, a normal part of aging or a true pathologic condition. They then tested a group of 38 men, all of whom complained of the symptoms of male menopause, including fatigue, irritability, and hot flashes; 32 of the men also reported impotence. The group was divided in two, those identified by the new test as having normal testicular functioning (15 of the 38) and those having abnormal functioning (23 of the 38). Heller and Myers administered testosterone to 9 men from the normal group and 20 from the abnormal, and found markedly different results. With abnormals, "definite improvement in the symptomatology was noted by the end of the second week in all of the 20 cases treated." With the normals, "none of the 9 patients demonstrated any improvement whatsoever," including sexual potence (although what was causing them to have such symptoms remained unclear). Heller and Myers concluded that male menopause did exist, that it was uncommon and it was eminently treatable. Recognizing that most practicing physicians would not be able to perform their complicated and invasive test, they recommended that clinicians give all their elderly male patients who complained of fatigue or impotency testosterone for two weeks. If they did not improve, the testosterone should be discontinued. If they did improve, it should be continued indefinitely.[71]

But others were far less comfortable with de Kruif's prescriptions, strenuously objecting to the idea of a male menopause and highly skeptical about the outcome data. A *JAMA* editorial on the "Climacteric in Aging Men" inveighed against both the concept and the practice of male hormone therapy. Comparing male to female menopause was inappropriate because men did not undergo sudden or abrupt changes. Moreover, investigators had not been able to confirm the accuracy of the Heller-Myers test, thereby leaving unresolved the question of whether a true hormonal deficiency existed among older men. Still, it remained uncertain whether opposition to testosterone use reflected insufficient data or medical conservatism. Other *JAMA* editorials, for example, noted that testosterone could "exert a tonic and stimulating effect" on patients regardless of their testicular levels.[72] And critics persisted in arguing that the hormone "may influence quite harmfully the physiologic and psychologic condition of previously well adjusted elderly men." It might encourage sexual behavior that would be "distracting and ill adapted to the needs of both husband and wife." There was also a concern about medicine sanctioning tonics and restoratives. A profession that was finally becoming scientific should not prescribe a putative "elixir of youth."[73]

The AMA's Council on Pharmacy and Chemistry was critical of the new testosterone compounds, although it, too, mixed scientific with social judgments. Since the marketed brands showed promise "in only a few conditions," the council would not include them in its roster of approved remedies.[74] It found that some drugs, such as those produced by Ciba, were inert, undercutting Ciba's claim that they cured impotence, senility, obesity, or the "climacteric of man." Other products were more effective in specific conditions. Treating eunuchs with testosterone produced "excellent results," although the council worried that since the therapy had to be continued indefinitely, "the economic factors are formidable." The drug also helped sexually underdeveloped young boys, but the council worried that it might lead to "masturbation and other undesirable behavior." It found no evidence of effectiveness in treating sterility and impotence and, most important, in treating "senility." Despite drug companies' eagerness to persuade physicians that "in the ampule of testosterone propionate lies the Foundation of Youth," reports of "attempted rejuvenation are not at all promising." The council was most comfortable with the stance adopted by James Hamilton, who feared the consequences of possible behavioral changes. "The patient experiences a sense of well-being on receiving testosterone. Moreover, euphoria is not uncommon and should be guarded against by strict insurance that the patient gets rest and does not overexert. Stimulation of an older man with androgens may cause him to feel younger and to attempt to lead the life of a younger man. The situation is to some extent like that of pouring new wine into old bottles."[75]

VI

The drug houses had no doubts about the efficacy of testosterone, determined to sell their versions of new wine. Schering was one of several companies to produce a lengthy and detailed "clinical guide" for physicians on male sex hormone therapy. Modeled on a medical text, it was organized by disease (sexual underdevelopment, impotence), replete with references to medical journals (227 footnotes), included exact recommended dosages, and tried throughout to encourage testosterone's use. The guide opened with a brief history of the hormone, starting with Berthold and Brown-Séquard, going on to Laqueur, Ruzicka, and Butenandt, and then brought the story to the present, when "primary male sex hormone is available for therapeutic use, its potency exactly known and invariable." Meticulous clinical research, "to which Schering has constantly contributed," now demonstrated testosterone's true efficacy. The company's preparations "represent the most advanced forms of male hormone therapy. . . . They are potent, chemically stable, and clinically dependable."[76]

Testosterone, the Schering claim went, acted not only on the genital tract but on "virtually every tissue." The Schering guide first attempted to establish the hormone's efficacy by documenting the successful treatment of sexual underdevelopment; "this condition offers a clear-cut instance of deficiency disease treated by wholly satisfactory replacement therapy." But sexual underdevelopment was too rare to constitute a substantial market. To make a profit, Schering had to convince doctors to prescribe the hormone as a general buffing tonic. Accordingly, it made the sweeping but dubious claim that "the male hormone increases resistance of the central nervous system against fatigue," and the equally unproven assertion that it relieves "nervousness, irritability, insomnia, increasing fatigue, apprehension, and restores effective mental ability not only in concentration but in the fulfilling of social and economic responsibility." Indeed, the company shrewdly linked these disparate symptoms to the new diagnosis—male climacteric—and promoted its product—male hormone—to remedy it. The guide explained that the male climacteric "syndrome," also known as "functional hypogonadism" (to give it more medical cachet), developed in middle-aged men, paralleling the appearance of menopause in women. Its symptoms represented "a "slackening of the life force . . . and, most conspicuously, a weakening or loss of sexual potency."

Schering had an antidote to this harrowing condition. Its testosterone product accomplished the "more or less complete rehabilitation of the patient," reversing all the symptoms, including impotence. Whatever failures marked the drug in the past were now irrelevant, for Schering's preparation was assuredly effective. It worked directly both on the tissues of the genital tract (restoring erections) and on the entirety of the constitution. The patient performed better because "the patient feels better." In fact, if the physician was uncertain whether his patient actually was suffering from testosterone efficiency, he should begin treatment with the hormone and see if it did any good.

Schering also claimed testosterone was effective against the benign enlargement of the prostate. It conceded that the mechanism by which it accomplished this shrinkage was unknown—perhaps the hormone improved muscle tone or reduced inflamation—but what mattered was that it relieved the symptoms of frequent urination and weak urinary flow, all the while heightening an overall sense of well-being. Testosterone was just what the doctor needed when seeing his older patients—and what the company needed to enlarge its profits.

What about side effects? Schering acknowledged the possibility of edema, particularly swelling in the ankles, and noted that a "sensitive" patient might experience stomach distress, but then the physician had only to reduce the dose or give sodium bicarbonate. In general, the preparation was "absorbed as a physiological substance and is not capable of any side-effects whatsoever in the adult male." The one major caveat was really more of a promotion for the drug than a caution: elderly patients taking testosterone might indulge in "undue activity," obviously of a sexual kind, which would strain their cardiovascular systems. What did the guide omit? Not a word appeared about testosterone's possible cancer-causing properties, the likelihood of reduced sperm count, or the threat of fluid retention to cardiac functioning.[77]

To increase sales, drug companies placed advertisements in medical journals that were even less subtle in their effort to promote physician enthusiasm. Ciba, Schering, and others regularly ran full-page advertisements in journals such as *Endocrinology*. The advertisements, like the guides, blurred any distinction between prescribing the hormones to combat a disease and to optimize well-being. The preparations apparently gave physicians near-magical powers. "Schering's potent endocrine preparations enable physicians to influence almost every phase of man's existence from the ovum through the span of life."

Many of the testosterone advertisements were heavy in text, listing all the symptoms and conditions that it might treat. Schering's presentation of Oreton promised it would provide "striking amelioration of the emotional and mental state." The drug was also appropriate for "Aging Men"—"Oreton has shown a definite tonic action in bringing about a sense of increased well being and renewed vigor." It also cured impotence (which in both younger and older men "frequently responds well to Oreton"). In all, Oreton would be "a highly effective means of 'finding' the man who is 'lost in his forties.' " It relieved the male climacteric that was "subtly" manifested in an "impairment of mental and physical energies," and worked equally well in cases of "profound emotional upsets, weakness, fatigue, and depression."[78]

Ciba, for its part, preferred to rely on eye-catching images. One advertisement for Perandren, its testosterone drug, devoted the top part of the page to a photograph of a grim-looking man wearing jacket and tie. The banner was in boldface type: "Psychic Trauma." The text itself explained that "the 6th decade often sees a conflict raging between the yet active mentality and the waning sexual forces." Perandren would effectively promote the sexual force. Another Ciba advertisement included a photograph of a very dejected man, dressed in tuxedo and black tie, staring vacantly into space. The banner: "The Fifth Age of Man." The text: "The male decline . . . is a period of sexual, particularly prostatic insufficiency . . . well recognized both by physical and mental aberrations. The rational therapeutic attack is . . . Perandren." Ciba's least subtle image was a drawing of a figure identified as Richard the Lion-Hearted, on a horse, helmeted and armored, thrusting his lengthy sword upward. The text, attributed to one Aretaeus the Cappadocian, compared a man who was "well braced in limbs, hairy, well voiced, spirited, strong to think and act" with another who was "shriveled" and "effeminate." Times had changed, the text explained. "Then there was no treatment. Now, there is Perandren." The shriveled could become erect.

VII

What place, then, did testosterone occupy within medicine at the outbreak of World War II? Despite some holdouts and critiques, the consensus, as evidenced by medical textbooks and journal overviews, shared a distinct bias toward its use. Among the most widely read and consulted texts was Cecil's, the work of Russell Cecil, professor of clinical medicine at Cornell medical school. His 1930 edition, which had appeared before testosterone had been synthesized, briefly addressed "Diseases of the Sex Glands," particularly sexual underdevelopment and aging. The "Treatment" section was very short, noting that "no active principle of the testes [has] yet been isolated" and that testicular transplants had a brief and uneven effect.[79]

The 1940 edition, with synthetic testosterone on the market, was lengthier, more factual, and far more optimistic about outcomes. Now Cecil's included several "before" and "after" close-up photographs of a eunuchoid's penis, demonstrating growth after testosterone administration. It also devoted an entire section to "The Male Climacteric." Although noting that it was not a common occurrence, "some waning of sexual activity occurs with advancing years," and along with it, as Brown-Séquard (very respectfully referenced in the text) and others had observed, "the fatigue and decreased vitality of advancing age." Without great elaboration or supporting data, Cecil's declared that "the male sex hormone has also been reported . . . to rejuvenate old men." Its bottom line was a hedge with a tilt to treatment. "While there probably is in some men a true climacteric which is relieved by the administration of testosterone propionate, the characteristics of this disorder and the details of its treatment are still to be worked out."[80]

Thus, a standard medical text instructed clinicians that testosterone was a reasonable if not yet proven treatment option. It was not snake oil and had very few adverse side effects.[81] Later editions of Cecil's included a section on "Complications," but the problems, such as edema, were not considered severe. The possibility that testosterone had cancer-causing effects received even less attention than in the case of estrogen. Cecil's advised that a patient who already had prostate cancer not be given male hormones. Since the hormones stimulated cell growth, it was "not desirable to employ them when malignant tendencies are known to be present."[82] But that did not mean that testosterone caused cancer, and no one should be concerned about prescribing it to cancer-free patients. The physician who believed that some of his elderly patients might be suffering from a "true climacteric" was well within the bounds of professional practice in dispensing the hormone.[83]

Then why did testosterone not become the male equivalent of HRT? With all these promotional engines running, why did it not rival estrogen? For one thing, men do not experience a dramatic menopause, a relatively fixed moment in time when a change of life is apparent. For another, men simply do not go to doctors as often as women do. They remain more distant from medical purview, whether the case involves a very specific ailment or a more general malaise. Third, it is possible that word of mouth worked against testosterone. Whatever the company claims for its use for erectile dysfunction, personal networks may have reported little to no success. (Compare this with contemporary conversations about Viagra.)

Finally, and perhaps most important, when men do visit physicians, they see a general practitioner, not a specialist. Unlike gynecology, there is no specialty devoted to male reproductive and sexual capacity. Urology, which emerged at the end of the nineteenth century, was essentially a surgical specialty, concerned more with physical problems than with performance or feelings. Urologists focused on the prostate and urethra, on incontinence, prostate disease, and venereal diseases. A few exceptions aside, they had never been drawn to the likes of a Steinach or Benjamin.[84] Thus, it would be highly unusual for a man complaining of fatigue or loss of libido to go to a urologist's office.

What would he be told were he to visit a general physician? In a man-to-man conversation, he would learn that testosterone might reduce his fatigue and improve his sense of well-being, and perhaps hear that it would not correct erectile dysfunction, impotence, or sterility. He would not likely be told of a risk of cancer, but he was certain to learn how expensive the drug was, far more so than estrogen. In 1943, Roche and Schering sold 25 milligrams of the compound to pharmacists for $6.25 or, adjusted to 1999 dollars, $60. Were a patient to receive injections three times a week in a doctor's office—the oral forms were less potent—the cost in today's dollars would be almost $900 a month.[85] In sum, it may have been weak demand that kept testosterone from becoming standard treatment. It was the patients, not the physicians, who kept testosterone from emulating estrogen.

VIII

Despite the weakness of consumer interest in testosterone and the lack of medical evidence to support claims of efficacy, researchers, anti-aging physicians, and pharmaceutical companies have not abandoned the drug. Testosterone still occupies a murky territory, with proponents encouraging its use, skeptics not taking a position, and opponents emphasizing its harmful side effects.

In the 1990s literature, as earlier, the most compelling argument for using testosterone in normal older men cites the results with abnormal younger men. The impulse to generalize from the pathological to the healthy continues to be irresistible. Journal articles promoting testosterone typically open with reports of how young men with a marked shortage of testosterone evince decreased sexual drive, inability to concentrate, reduced muscle mass, and loss of strength and stamina; once they receive testosterone, they experience improved libido, better mood, and increased muscle mass and greater stamina. Two assertions then move testosterone treatment from the young to the old. First, noting that testosterone levels in older men average 50 percent to 75 percent lower than in younger men, proponents maintain that "there is no reason to think that the tissues of older men require less [testosterone] than those of young men." Second, they justify interventions on the well-trod grounds that symptoms of aging resemble the symptoms of sexual underdevelopment. Thus, one geriatric center in St. Louis used a questionnaire to learn whether patients were suffering from "low testosterone syndrome." "Are you sad or grumpy? Are your erections less strong? Have you noticed a recent deterioration in your ability to play sports?" Were the answers yes, testosterone should be given.[86]

But even within this framework, prescribing the hormone is problematic. In the overwhelming majority of older men, testosterone levels are in the "low but normal" range for younger men. Someone with these scores who was in his twenties or thirties would not be diagnosed as "sexually underdeveloped." So too, almost all older men have approximately the same levels of testosterone, making it more difficult (but not impossible) to define a common condition as pathological. Others respond by casting doubt on the accuracy of the laboratory measurements. Joyce Tenover of Emory's School of Medicine contends that "the definitive diagnosis of testosterone deficiency . . . for this age group has not been established." In one type of test, only 5 percent of men over sixty had levels altogether below the normal range for younger men. But in a second type of test, 11 percent to 36 percent were testosterone-deficient, and using a third technique, "as many as 50 percent of men aged more than 60 years could be testosterone deficient." But then how should a clinician or patient proceed? Assume the best or assume the worst? Laboratory-hop until one has the desired finding, confirming or disconfirming a need for a testosterone supplement?[87]

An even more basic question is whether older men actually benefit from receiving testosterone. Some investigators report that normal elderly men given testosterone do demonstrate an increase in muscle mass, a loss of body fat, and, according to some studies, increased strength as measured by a hand grip. Others find improvements in mood. Still others cite studies in mice that point to improved memory, suggesting "a potentially important role for testosterone in modulating age-related cognitive decline." But critics discount these findings. Although muscle mass increased, there is no evidence that testosterone actually improved day-to-day performance; recipients were not able to walk farther, or quicker, or carry heavier packages, or carry out any one of life's daily functions better than those not receiving testosterone. So too, although testosterone did appear to promote "some aspects of sexual arousability," that is, men receiving it thought more about sex, sexual performance was unaffected.[88]

Even physicians who believe that older men should receive testosterone concede a paucity of supporting data. Tenover calculates that "the total number of older men treated in all published studies combined is somewhat less than 75, and the length of therapy has varied from 1 to 18 months."[89] Others concede the weak design of the studies. Selection criteria differ on what precisely constitutes a testosterone deficiency; the means of administering the hormone are not the same—some researchers inject the hormone, others use a skin patch. There is not even agreement on what testosterone levels are needed to produce any benefits.

However dubious the gains, the risks are not trivial. The major concern has been prostate cancer. Testosterone's relationship to prostatic tissue is like estrogen's relationship to uterine tissue—both stimulate growth and, therefore, potentially, cancer cell growth. Indeed, one treatment for prostate cancer is chemical castration, used expressly to reduce testosterone levels. Does testosterone itself cause prostate cancer? No one is certain. Most of the studies are too short-term. Prostate cancer cells are slow-growing, often taking months and years to spread, and so to conclude that elderly men in a three-month trial of testosterone did not develop prostate cancer is to say almost nothing. What about such surrogate markers as the level of prostate-specific antigens (PSA)? Again, the evidence is inconclusive. One study did find elevated PSA levels in subjects taking testosterone after three months (92 percent of 13 subjects); and even after the testosterone was stopped, the levels did not drop back. (It is worth noting that proponents of testosterone more often cite this particular study for its findings that muscle mass increased than that PSA scores climbed.)

Given the stimulating effect of testosterone on cell growth and the weakness of the data on benefits, the medical literature is replete with advice to be cautious in dispensing testosterone. "Supplementation in older men," Tenover concludes, "can not be currently recommended, and more data are needed . . . before clinical decisions regarding T therapy can be rationally based." As would be expected from the estrogen story, calls for more research continue. "Larger and longer term studies . . . [are needed] to determine both the risks and benefits." But for now, the case for "enhanced survival and improved function" through testosterone has not been demonstrated.[90]

Given the state of the data and the principle of do no harm, one might have anticipated that practicing physicians would not be prescribing testosterone for healthy older men. That is not the case. The drug, which is FDA approved for testicular injuries or malfunction, is dispensed "off label" to older men, and rough estimates are that its use has doubled in recent years. The drug companies do their best to promote it. FDA rules prohibit them from advertising or marketing testosterone as an anti-aging therapy but there is ample room for improvisation. The Web site for Androderm notes that men over fifty experience a gradual reduction in testosterone and then purposely links the treatment of sexual underdevelopment to aging. Observing that "men with age-related testosterone deficiency represent the largest group of hypogonadal men . . . it is reasonable to wonder if [testosterone] can be used to halt or reverse some problems of male aging." It hedges on the question of "whether symptoms . . . are a normal part of getting older or should be considered evidence of a medical condition," but leaves no doubt that physicians would be wise to write a prescription for testosterone.

The media promotes testosterone as well. Although the tone of many newspaper stories on testosterone is cautious, the press cannot resist capturing reader attention through a notable case of success. The *Cleveland Plain Dealer* reports that one man who had lost his desire to "kick butt" went on the Internet, found a doctor willing to prescribe testosterone, and is now enjoying a "new life." *USA Today* tells about a man who suffered from fatigue, hot flashes, and mood swings; after taking testosterone, he jogs six miles and has an active sex life. There is another man whose memory improved, another who took testosterone and was inspired to start a new business, and still another who "has more energy, can stay up late, and, as he puts it, does 'pretty well in the sack.' "[91]

The so-called anti-aging clinics make testosterone one of their staples. The Life Extension Institute promotes testosterone for a lean body mass, and uses a Charles Atlas–like figure as the logo for its pages explaining how it works. Web sites pitch and provide. SmithKline set up the Web site testosteronesource.com, where visitors could raise their Awareness and get an Education about the drug, read Patient Stories, learn about their Options, Ask the Expert, and take advantage of the "Physician Referral Service" to obtain the names of doctors who treat "testosterone deficiency." Enter testosterone on one search engine and Cenegenics, dedicated to "youthful aging" (whatever that means), comes up first. It will schedule a free physician consultation and evaluation: all you need do is call 888-Younger and give them your name, e-mail address, telephone number, and (for some unstated reason) tell them whether your annual household income is below $50,000, between $50,000 and $100,000, or above $100,000. Of course, there is no shortage of Web sites providing herbal alternatives to testosterone, including Testron, that will maximize virility, potency, and serve as a "natural aphrodisiac."[92]

In sum, testosterone is easily available to anyone who wants it. There are physicians in private practice or in anti-aging clinics or available on-line or by telephone for a quick "consultation," who will write the prescription. They face no reprimands from medical societies or state boards of professional conduct or the FDA. If consumers believe that aging is a form of sexual underdevelopment, they will have no trouble obtaining the hormone. In one sense, their use of testosterone will be following an old tradition that began with Brown-Séquard. In another sense, they will be following a new path, giving us a glimpse of how future enhancements will enter the market. If consumers confront profound medical ambiguities about a substance that in one form or another has been around for more than one hundred years, imagine what they will face when new genetic techniques and novel drugs promise to deliver extraordinary advantages.

SEVEN

The Price of Growth

Nowhere have the goals of cure and enhancement become more entwined and calculations of risks and benefits more salient than in the use of growth hormone for children of short stature. The story of growth hormone differs in details from estrogen and testosterone, but what is most remarkable is the many similarities. Shortness has turned into a medical condition, just like aging, and physicians have taken as their duty making children and parents happy. Drug company promotions of growth hormone make their efforts on behalf of estrogen and testosterone seem tame. Indeed, no experience better illustrates how the forces of science, medicine, culture, and commerce have combined, and will combine, to create and popularize enhancement technologies.

I

Although physiologists in the early 1900s already appreciated that growth hormone (secreted by the anterior portion of the pituitary gland) played a critical role in determining physical size, the state of their knowledge was too rudimentary to design therapeutic or enhancement interventions. Surgeons might occasionally excise a tumor from a patient's pituitary gland so as to prevent giantism, and some physicians prescribed pituitary extracts, notwithstanding their dubious quality, to try to stimulate growth. But medical interventions were generally unable to affect abnormalities in body height.

The situation remained practically unchanged into the 1950s. Endo-crinologists were better able to distinguish among a variety of children with short stature, or "stunted growth," as they called it. They recog-nized that the great majority were merely experiencing a delay in development; they would catch up to their peers in adolescence. Endo-crinologists also recognized that a small percentage of short children were "primordial dwarfs," their short stature one of several physical abnormalities that were probably genetic in origin. But they had a third category that interested them most, "pituitary dwarfs," children who were suffering from a deficiency in the hormone. They had been normal in size and appearance at birth and for months afterward; but then between the ages of one and four, their growth rates lagged. In some instances, a diagnostic workup uncovered a pituitary tumor that was impeding the production of the growth hormone. But most cases were idiopathic, of unknown origin. These children thrived in all other ways. They were not sickly—they just were not growing. Given the crude state of laboratory testing, it was very difficult to learn whether they were truly hormone deficient. The pituitary gland's excretions were irregular (more often discharged at night than during the day) and blood analysis of hormones was indeterminate.[1] Thus, pituitary dwarf-ism was a judgment call. If the child was several standard deviations below normal height for his or her age group, if the growth curve remained flat, and the levels of pituitary hormone appeared to be low, then the problem might well be hormone deficiency.[2] Yet even after giving the diagnosis, physicians could not do much to ameliorate the condition.

Over the 1950s, investigators explored various possible sources for growth hormone. Researchers from the fields of endocrinology, physiology, and biochemistry did manage to isolate growth hormone (somatotropin) in pure form from the pituitary glands of cattle, but administering the substance to the children brought no changes. Lilly and Merck were actively collecting kidneys from rhesus monkeys in order to cultivate polio viruses for use in the Salk vaccine, and they were willing to retrieve pituitary glands as well. Researchers hoped that growth hormone isolated from monkeys, which was chemically differ-ent from growth hormone taken from cattle, might be the answer. But it, too, demonstrated no efficacy in humans.[3] Several pharmaceutical houses did market "endocrine preparations," but the medical consensus was that they were "without effect in achieving growth in children."[4] Worse, endocrinologists feared that the preparations might have the "adverse effect of actually inhibiting endogenous growth hormone production that the child may have."[5]

The ideal substance for treatment, of course, was human growth hormone. Although it was exceptionally difficult to obtain, the little that was available seemed highly effective. In 1958, M. S. Raben, a professor at Tufts Medical School, reported that he had injected a seventeen-year-old pituitary dwarf with human growth hormone (HGH) over a period of ten months, and the boy grew at a rate of 2.6 inches a year (as compared to a previous growth rate of 0.5 inch).[6] Supply, however, could not approach need. "The scarcity of the raw material," as one review noted, "has sharply limited the use of this preparation in the deserving patient."[7] To obtain human growth hormone, it was necessary first to collect pituitary glands from cadavers. That required contacting families of recently deceased patients and persuading them to give permission for an autopsy—the families were generally not informed that the purpose of the procedure was not to learn the cause of death but to remove the pituitary gland. (If they knew the truth, it was feared, they would say no.) Moreover, interns and residents no longer routinely asked families about autopsies, and even when they did, families generally refused. In the odd case when permission was granted, the extraction process yielded only about 1 to 2 milligrams of growth hormone per gland.[8] Since the treatment regimen for pituitary dwarfs called for 1 to 2 milligrams a day, three days a week for twelve to twenty-four months, human growth hormone was, for all intents and purposes, unavailable to the children it might have helped.[9]

The attempt to substitute other hormones for human growth hormone also proved futile. Some endocrinologists hoped that anabolic steroids, which increased muscle bulk and weight, might increase height, but that proved illusory. There was an expectation that extracts from thyroid glands would exert a positive effect, but that too failed.[10] Facing this desperate situation, in 1963, the National Institutes of Health, through its National Institute of Arthritis and Metabolic Diseases (NIAMD), took the extraordinary step of funding a National

Pituitary Agency (NPA) at Johns Hopkins. Its mission was to expand and improve the system of gland retrieval and to learn as much as possible about the uses and properties of human growth hormone.[11] The NIH also hoped, though it kept this to itself, that the NPA would prevent desperate parents from fueling a black market in growth hormone.[12]

In order to increase donation rates of pituitary glands, the NPA decided to issue a bimonthly newsletter, a far more controversial step then than now (when drug company advertisements to consumers and organ donation appeals are commonplace). The formal announcement of the newsletter was remarkably defensive in tone, apologizing for this exercise in "lay publicity." It conceded that calling attention to the efficacy and need for pituitary growth hormone might be misconstrued as exaggerating its potency and encourage quackery. It also appreciated that clinicians were worried that spreading the word about the effectiveness of human growth hormone would cause them to be "overburdened by patients in need of specialized help." "It is morally reprehensible," the NPA itself declared, "to engender hope within a suffering patient when this hope can not be realized." Nevertheless, it proceeded with the campaign. "We are faced with the reality that unless people contribute their pituitaries, not only will we not have the hormone needed for present investigative therapy, but we will not have the hormone necessary to establish its structure so that it may be eventually synthesized and made readily available."[13] In addition, the NPA encouraged magazine articles, television interviews, and films about its work. It even attempted, without success, to insert pituitary dwarfism and the need for cadaveric glands into the plots of such popular television medical shows as *Ben Casey* and *Dr. Kildare*.

The NPA tried to educate hospital interns and residents about the importance of requesting autopsies. Its very eagerness not only led it to ignore principles of informed consent but also to engage in immigrant bashing. The NPA blamed the shortage of autopsies on the "large number of interns and residents from foreign countries now working in American hospitals." Apparently, they had "never been exposed to the fine purpose and established practices of autopsies," and were ignorant of the need for a "final post-mortem investigation of all patients."

Accordingly, the NPA commissioned a thirty-minute color film to imbue this "Anglo-American tradition."

By dint of all these efforts, the NPA at first collected some 50,000 pituitary glands a year, happy to pay a two-dollar fee for each. From July 1, 1964, to January 31, 1965, it sent some 30,000 milligrams of the hormone to investigators, enough to use on some 400 patients for a period of up to three months.[14] By 1970, its totals reached 70,000 milligrams. Even so, supplies were very tight and pituitary hormone went only to selected investigators. "The objective has always been basically research," admitted NPA's advisory board, "even though some of the literature used in the promotion campaign to collect pituitary glands may have permitted other interpretations."[15] The closest the NPA came to supporting clinical practice was to give a team at the University of Buffalo growth hormone for a limited number of patients in order to establish clinical guidelines.

Thus the only way a child could get growth hormone was by enrolling in a research protocol—which lent an element of coercion to subject recruitment even if the practice could be justified by the scarcity of the substance and the many unknowns about its effects. Enrollment criteria were strict: eligible children had to have a growth rate in the lowest 25 percentile of their age group, and measure below 5 on a standard laboratory hormone test. Subjects who completed the research regimen were eligible to continue to receive growth hormone afterward and at no charge, but because of limited supply, the hormone was provided only eight months a year and stopped completely once the child reached 5 feet.[16] Although not so intended, the system of distribution as well as cultural expectations about size and gender gave a distinct advantage to white males. As one survey found, the NPA cohort was 68 percent male, 32 percent female; the average age at which treatment began was 9.6 years, and the average duration of treatment was 2.9 years. Not only girls but also blacks were significantly underrepresented among the research subjects—in fact, the girls and blacks who were enrolled were considerably shorter than the white boys in the protocol.[17]

In 1965, when the annual Ross Conference on Pediatric Research brought investigators together to discuss human growth hormone, the principal conclusion was how much still remained to be learned.

Everyone agreed that only human growth hormone offered pituitary-deficient children "the opportunity to attain a reasonably normal adult stature." Yet, no consensus existed on what constituted the most effective dose. The current regimen was 2 milligrams daily, three times a week. Might twice as much twice as often work better? No one knew. "We have no information on what dose would be optimum."[18]

Over the next twenty years, the NPA increased the retrieval rate of pituitary glands and improved methods for extracting growth hormone from the glands. With a greater amount of the hormone available, former research subjects were able to receive it twelve months a year, and the treatment cap was raised to 5 feet 6 inches for boys, and 5 feet 3 inches for girls. Even so, the hormone remained in very short supply and the NPA still rationed its use.

The frustration in knowing how to cure a condition but lacking the means to do so was unusual in post–World War II American medicine, but not unprecedented. The closest parallel was end-stage kidney disease, where the ability to save lives was limited by the number of dialysis machines available. To cope with dialysis crisis, a number of medical centers established "Who Shall Live?" committees to allocate the scarce resource among would-be recipients. Nothing like that occurred with growth hormone. Since the condition was not life-threatening, NPA researchers were left to their own devices.

Indeed, with medicine unable to satisfy demand for the hormone, psychiatry entered both to evaluate potential candidates and to treat children who were unable to get the hormone. Over the 1960s and 1970s, psychiatric research on the consequences of short stature indicated that parents, teachers, and strangers tended to react more to the physical size of a child than to his or her chronological age, so that short children experienced a "babying" effect. Because of this infantilization, they were prone to developmental deficiencies, immature, and lacking self-confidence.[19] Even health professionals who underwent sensitivity training "continued to find themselves bending down, talking in an unsophisticated manner, and expecting less mature responses from a 9- or 10-year-old child who looked like a kindergarten child."[20] As for the child, "when one's sense of self-worth and personal identity are established mainly on the basis of the reactions of others, major differences in appearance may lead to enduring and profound personality difficulties." These children often lived "in a secluded inner world of intensified feelings, sentiments, and emotions," their personality traits "mainly a result of experiences associated with dwarfism." They tended to be "less aggressive, less excitable, less dominant" than those of normal height.[21] They suffered from "low self-esteem, social isolation, low level of aggression, and affective withdrawal."[22] They wavered between "feelings of helplessness and hopelessness."[23]

The combination of endocrinology identifying a state of hormone deficiency and psychiatry analyzing the degree of maladjustment suggested that shortness of stature was a disease, and by no means a trivial one. The very short child was a sick child. And yet, at the very same time, physicians were uncomfortable with a definition of an otherwise healthy child as sick. Smallness of size did not easily fit into a disease category. The tension between the two perspectives helps to explain why, well before growth hormone became readily available, physicians were already anticipating a very special dilemma. They wondered if, in the future, parents would be demanding that their children of normal height be made taller. Like the authors of the 1930s high school texts that asked students to think about the ethics of making normal children taller, the physicians, too, wondered whether cure and enhancement would remain separate. Thus, in 1976, Alfred Bongiovanni, a leading pediatric endocrinologist, deeply regretted that growth hormone was not available for pituitary dwarfs. "There is a dire need for ample supplies of HGH and there is an obligation to provide public and private funds to achieve this goal in the near future." But then he added: "It may be predicted that in the wake of such success these pages will carry a commentary, 'The Use and Abuse of HGH,' containing an account of the hazards of large doses and proscriptions against its use in normal somewhat small children." Scarcity bred difficulties, but abundance was likely to bring its own headaches.[24]

II

And then suddenly the scarcity of human growth hormone and the small number of children who had received it became a source of great relief. In April 1985, the Food and Drug Administration announced that three recipients of human growth hormone had died of Creutzfeldt-Jakob (CJ) disease, now known as prion disease. Early in March 1985, a Stanford University pediatric endocrinologist, Raymond Hintz, had written the FDA that a twenty-year-old man whom he had treated with human growth hormone over a fourteen-year period had died from CJ disease. Hintz wondered whether growth hormone was the source of the infection: "The possibility that this was a factor in his getting Creutzfeldt-Jakob disease should be considered."[25] The disease was not only rare (one case in a million) and usually affected only the very elderly, but also growth hormone seemed a possible vector of transmission. Since the substance was derived from pooled cadaveric pituitary glands (usually about 15,000 of them), the glands might easily have included infectious tissues from one or more patients not suspected of having CJ disease. Although the pituitary extracts were filtered, not much was known about the efficacy of sterilization against this strange agent.[26] In response, the NIH quickly convened a meeting of experts. They recommended a compromise: a halt to all research with human growth hormone, but a continuation of therapeutic use. They reasoned that since the time window for treatment with growth hormone was limited—once the child reached adolescence the substance would no longer have any effect—and there was only one case with no definite link to growth hormone, the risks of transmission appeared less than the benefits of growth.

News of the Stanford case and the experts' recommendation provoked a fierce debate among pediatric endocrinologists. Was the one fatality a coincidence or was it actually due to growth hormone? Was the risk of CJ disease great enough to halt therapies to several thousand children? The controversy ended abruptly a month later when two more cases of CJ disease appeared among growth hormone recipients, one a thirty-two-year-old man in Dallas, the other a twenty-two-year-old man in Buffalo. At this point, the NIH halted all distribution of growth hormone.[27]

Excruciating questions arose immediately. Just how grave was the risk to those who had already received growth hormone, not only in the United States but also in England and Europe as well? Were many

batches of the substance contaminated or only a few? Were the earliest
growth hormone products more dangerous than later ones? And where
did responsibility lie for the outbreak? Could it have been averted?
What measures might have been taken to better sterilize the product?
Finally, physicians had not advised parents about the possibilities of
such a side effect. Why had they ignored the risk? Parents might not
have chosen to treat for height had they received information about the
possibility of the treatment transmitting a fatal disease. (In a very short
time, another group of parents, those with children who had hemo-
philia, would be asking the same questions about fractionated blood
plasma, but they at least had the advantage of knowing that the intervention was lifesaving.)

Answers came slowly and with ambiguities. The NIH assembled a
team of investigators to survey the 6,284 recipients of growth hormone
in its pituitary program. Even as the team went about its work over the
period 1985 to 1991, physicians reported two more cases and the team
itself turned up an additional two, bringing the total deaths from CJ disease linked to growth hormone in the United States to seven. Although
the victims resembled the larger group of growth hormone recipients,
they had two distinguishing features: they had begun treatment prior to
1970 and they had remained in treatment longer than average. The team
could not pinpoint specific lots of hormone or batches of pituitary
glands responsible for the outbreak. Two specific lots were common
to the seven infected children, but other children had also received
growth hormone from them and had not, at least not yet, contracted CJ
disease.

What risk did children who had received growth hormone face?[28]
There were no certain answers. CJ disease has exceptionally long
latency periods (ten to twenty years), and many recipients were still
well short of that benchmark figure. The optimistic view was that the
seven fatalities had the bad luck of consuming "contaminated hormone
preparations [with] a small number of infectious units scattered ran-
domly among individual vials of hormone." But it was entirely possible
that the number of cases would mount over time. In England, for example, with about the same number of fatal cases identified, investigators
were warning that "all patients treated [with growth hormone] in the

UK must, at this stage, be considered as at risk."[29] By 1994, almost ten years after the initial outbreak, the number of deaths had climbed, but not to catastrophic levels: 51 cases of CJ disease worldwide attributable to growth hormone; 10 of them were in the United States, 10 in Britain, and 25 in France. By 2000, the United States had a total of 21 confirmed cases; France, 62 cases; England, 32 cases. But even now, no one can be absolutely certain that the outbreak is over.[30]

The answers to the questions of responsibility for the outbreak were also equivocal. With the benefit of hindsight, prudent investigators might have anticipated the risks of growth hormone extracts before 1985. However, knowledge about CJ disease had never seemed particularly relevant to their work. The infectious character of CJ disease was established in 1967–1968, mostly through the research of D. Carleton Gajdusek. Working closely with anthropologists, he identified kuru, as it appeared in the Fore tribe in New Guinea, as a form of CJ disease and attributed its spread to a mourning ritual in which women and children of the tribe ate the brains of the dead soul. In 1969, Gajdusek and a colleague reported in *Science* that chimpanzees inoculated with tissue from patients diagnosed with CJ disease themselves contracted it, thereby confirming its infectious character.[31] But kuru seemed an exotic disease, irrelevant to developed countries. Surely no one there would come into such intimate contact with the tissue of a person who died of kuru or of CJ disease.

This seemingly reasonable expectation was soon proved wrong. In 1974, the first known case of CJ disease transmitted person to person in the developed world appeared in a report of a corneal transplant. The donor, it turned out, had died of CJ disease and the recipient then contracted it. In 1977, another case of surgical transmission came to light, this one through the instruments that had been used on a CJ patient and then sterilized in a solution of alcohol and formaldehyde. The standard solution did not kill the agent. These instances notwithstanding, other specialists, including pediatric endocrinologists, failed to see a relevance to their own practices.

There was an exception. In late 1976, an English veterinary scientist wrote to the British Medical Research Council (MRC) to warn of the possibility that sheep scrapie-like agents, akin to the CJ agent, might be transmitted through growth hormone. It took the MRC more than a year to begin to investigate his warning. (Indeed, it took a lawsuit by parents of children infected by CJ disease to get this material on record.) One MRC consultant had taken the threat of contagion seriously: "We are in the uncomfortable position of suspecting the worst but not knowing how bad the worst is. Any clinician who uses growth hormone must be made aware of the gruesome possibilities."[32] But no steps were taken to make them aware. The risk inherent in using cadaver-derived human growth hormone was never communicated to clinicians in England, or in the United States. What knowledge there was, was scattered among neurologists, anthropologists, and veterinarians, but not shared among pediatricians, endocrinologists, or growth specialists. If they thought of CJ at all, it was as an elderly person's degenerative disease.

When all the facts became known, the children who received growth hormone had to live with the fears of succumbing to a deadly disease, and their parents had to live with the guilt of putting them at mortal risk because of something as minor (compared to premature death) as short stature. The rest of us have to confront the stubborn question of how to calculate the risk of a procedure when almost all the experts see no risk. Right up to the outbreak, the medical literature confidently asserted that "growth hormone has been shown to be safe in patients deficient in the hormone."[33] The confidence was misplaced in this case—might it be misplaced in another? Absent acute disease but present unhappiness or the prospect of enhancing performance and capacity, just how much risk is one prepared to take?

III

In 1985, Genentech, a California biotech company whose synthetic version of growth hormone had been undergoing clinical trials for several years, scored an extraordinary coup when CJ disease was linked to cadaveric sources. As one NIH investigator concluded, when "the hormone was officially executed . . . only Genentech was not in mourning."[34] Even had the company been forced to compete with the natural product, it would have profited handsomely. But by the stroke of (mis)-fortune, it had the territory almost entirely to itself.

Just how Genentech managed to produce a synthetic version of growth hormone was the subject of a long-running lawsuit between the University of California at San Francisco (UCSF) and Genentech. The company itself was founded in 1978 by Robert Swanson, a brilliant venture capitalist, and Herbert Boyer, a world-class genetics researcher at UCSF. Both recognized that the new tools of DNA research—which allowed for the isolation and replication of particular genes—had the potential to revolutionize the creation of pharmaceutical agents. Just as the new company was getting organized, investigators at UCSF, including Peter Seeburg, managed to isolate the gene responsible for producing growth hormone. Genentech then recruited Seeburg to continue his research under its auspices; UCSF, for its part, had Seeburg sign an agreement that gave it all property rights to the work he had already performed in the university laboratory.

What happened next is the subject of bitter dispute. According to Seeburg, he and his Genentech colleagues, including David Goeddel, were unable to replicate their earlier work and isolate the growth hormone gene for Genentech; after several frustrating failures, Seeburg raided his old UCSF lab, stole the gene, and he and Goeddel swore never to reveal the theft. Goeddel claims that the story is pure fabrication, that he was able to isolate the gene and, in fact, published the results in *Science*. But Genentech did not back up Goeddel's story, and his research notebooks lacked the specific data entries that would have confirmed it. The question of theft is unlikely ever to be resolved because the lawsuit between Genentech and UCSF was settled in November 1999, when Genentech agreed to pay UCSF $200 million. At the university's suggestion, $50 million of the $200 million was to be used to construct a science building at its new research campus, and Genentech would have the right to name it. The project was completed in 2003, and the building opened as Genentech Hall.[35]

By whatever route the gene arrived, Genentech figured out how to insert it into *E. coli* organisms and produce large quantities of growth hormone. In October 1985, the FDA, in record time, only four months after Genentech's application, approved the bioengineered growth hormone Protropin. Everyone who had been receiving pituitary-derived growth hormone now had an effective, albeit expensive, substitute available. (Genentech set the price high—$18,000 for a year's supply.) Even so, the Genentech feat meant that the need to ration the substance was over. An intervention that had once been managed by a federally funded not-for-profit organization was under the control of a for-profit and exceptionally hard-driving drug company. Care and treatment would no longer be controlled by a small team of endocrine experts. For better or for worse, clinical investigators, endocrinologists, physicians, parents, children, and bioethicists, along with drug companies, a U.S. attorney, and a federal judge would shape the growth hormone story.

Looking back on the pre-1985 period, one endocrinologist who had closely followed the uses and abuses of Genentech's growth hormone pined for the not-so-good old days. Yes, the drug had been in short supply, but at least no hard choices had been required. The little growth hormone that was available was reserved for children who were demonstrably growth hormone deficient. But once supply was assured, a host of questions surfaced. First and foremost, who should receive the hormone and under what regimen? Clearly, children who were growth hormone deficient, which was the use for which the FDA had approved the drug. But what about short children who were not clearly hormone deficient by the old and rigorous measurement standards? Was the traditional scoring scale for hormone deficiency too restrictive? Moreover, it was still unclear at what age treatment should be started and at what age it should be stopped. When the hormone had to be rationed, it made sense to begin as late as possible and end as quickly as possible. But was that right? Finally, with the shortage over, it was necessary to revisit the question of optimal doses and frequency.

The lack of clinical data made each of these questions difficult to answer. Yes, growth hormone should go to children with growth hormone deficiencies, but methods for measuring its levels were still unreliable. Since the pituitary gland released growth hormone episodically, no single blood draw provided an accurate reading. But to take blood every fifteen minutes over a twelve-hour period at night was difficult to arrange, expensive, and burdensome to the child. A more preferred method was to inject the child with a substance known to stimulate the output of growth hormone (such as insulin) and then immediately measure the amount of hormone in the blood. But the assays for identifying growth hormone in the blood were not very accurate—different laboratories used different assays of varying quality—and the margins of error were considerable. In light of these problems, some endocrinologists recommended that the better part of wisdom now was to forgo hormone measurements altogether and rely instead on diagnosis by trial (as had been suggested for testosterone). Administer growth hormone to very short children and see what happens.[36] If the intervention produced results, keep using it. If not, discontinue.

To the extent that reliance on hormone measurement persisted (which had more to do with insurance company rules governing reimbursement than medical judgment), the criteria for determining hormone deficiency relaxed and the gap between a normal and abnormal reading narrowed.[37] Thus, the definition of growth hormone deficiency varied not by medical criteria but by supply.[38] "Category creep" was under way.

Some data drawn from the NPA research suggested that growth hormone administration should begin at a young age, should be given six or seven days a week, and the dosage should increase as the child approached puberty. But the findings were soft; the number of clinical trials was low, as was the sample size, and there were considerable differences among the research subjects in terms of age and level of hormone deficiency. Nor was there agreement on when to stop treatment. Some pediatric endocrinologists relied on X rays of bone formation to indicate when growth was ending, but others thought the method was imprecise. Some stopped treatment when the child completed sexual development but others believed that this was too early. Still others stopped growth hormone when the child was one or two standard deviations below normal height, or had reached the average height of the two parents combined, but critics found these end points too arbitrary. As a result, variation became the rule, and endocrinologists were free to follow their own instincts.

The most glaring gap in the data was a lack of solid evidence on outcomes. The most obvious question was still unanswered: What return in inches will come to a hormone deficient child who takes growth hormone? Or as some parents might ask it: How many inches will my child grow if I make the $18,000 investment in growth hormone? The very diversity of regimens, the small number of trials, the absence of randomization, the lack of agreement on stopping points, and the number of children "lost to follow-up" because they stopped seeing their pediatricians precluded conclusive findings.[39] One major review of findings (in 1991) looked at the leading six studies but could not come up with useful guidelines. None of the trials had used the gold standard of a placebo control group, so there was no way to know how much growth would have occurred anyway. The range of age at which treatment began varied from eleven to fourteen; the duration of treatment varied from four to six years. Only three of the studies included information on the subjects' height before and after treatment. With all these limitations in mind, it seemed that three of the studies showed an average increase in height (from 121 to 157 centimeters). All reported an improvement in subjects' standard deviation from normal height; they were on average −4 when starting treatment but −2 when concluding treatment. But no group receiving the hormone achieved average height.[40] So, in principle, growth hormone benefited very short children. But just how well it worked, precisely for whom, and under what conditions was not known.[41]

IV

All these uncertainties about treatment would not have been so consequential had the only recipients of growth hormone been dwarflike children. Their handicap was so severe as to justify diagnosis by treatment or an intervention that had a good but not exceptional prospect of success. But from the moment that researchers began to appreciate the role of hormones in stimulating growth, everyone, physicians and the public alike, had anticipated the likelihood of "diagnostic creep," that some number of short but not hormone deficient children would seek treatment. In the period of scarcity before 1985, this possibility was theoretical. But with the hormone readily available, the issue became real, and the debate acrimonious.

Some pediatricians and endocrinologists were uncompromising in opposing growth hormone for short but not hormone deficient children.[42] In part, it reflected their acute skepticism about its efficacy. Data on whether growth hormone actually made non–hormone deficient children grow taller was, at best, equivocal. They were also skeptical of the extent of the burden that short children carried. "Neither we nor others," contended one group of researchers, "have been able to document markers of adverse psychological effects in normal short children."[43] Indeed, the negative psychological and social effects of a daily routine of injections might be even more damaging.[44] They were concerned, as well, in the aftermath of CJ disease, about new and unanticipated risks. The recombinant growth hormone product was free of infectious agents, but it might render children carbohydrate intolerant, stimulate an overgrowth of tissue and bone, cause carpal tunnel syndrome, or increase the risk of tumor formation and tumor growth. None of these effects were as yet well documented, but that did not mean that the intervention was risk-free.

Other objections addressed matters of equity. Critics worried that if growth hormone for normal short children became standard treatment, an undue amount of health care resources, particularly federal benefits, would be devoted to it. Growth hormone was so expensive that far more urgent public health needs would be neglected. The numbers were easy to do. An estimated 90,000 children were in the lowest percentiles of height; were they all given growth hormone at federal expense, the national bill would amount to over $8 billion. The alternative to a government subsidy was equally unacceptable. To allow wealthier families to pay for it themselves and leave the poor to their own devices would only entrench inequities in health care. We might reach a point wherein only poor children were short children.[45] And even were growth hormone effective and made widely available, all that would follow would be a shift in the distribution curve. Children who had been one step above the lowest percentiles in height would now drop to the bottom; they would then seek treatment, and the reshuffling process would begin all over again.

Opponents were also adamant in insisting that medicine should resist crossing the line from cure to enhancement. Shortness was not a disease.

Some considered it a "natural human variation." Others thought of it more in terms of stigma, regretting the prejudice but not finding it catastrophic.[46] As a group of University of Chicago endocrinologists put it: the intervention would constitute "a form of cosmetic therapy, rather than a treatment for a disease." Not that all cosmetic therapy should be ruled out. This one team would allow cosmetic surgery for a cleft lip because the condition was "so disforming." But treating shortness, like altering an unshapely nose, it found "unacceptable," representing an "unwarranted tampering with nature." It never did explain what made one condition disforming and another merely annoying, or when medicine should tamper with nature and when it should not. But the team worried that if endocrinologists ignored their advice, the specialty would confront the nightmare of giving growth hormone to a "child of normal height to make him or her a better basketball player."[47]

Equally credentialed and every bit as aggressive, proponents defended giving growth hormone to what they labeled children with idiopathic short stature. (In medicine, as in other disciplines, labels count for a lot.) These were not short normal children but patients who evidenced the clear symptoms of a yet unidentified pathology. They objected strenuously to basing decisions on laboratory-derived growth hormonal levels. The measurements were far too unreliable. Why enshrine as the gold standard a diagnostic procedure that was, at best, brass?[48] Indeed, many American endocrinologists had already given up testing hormone levels, and in countries such as Australia, physicians did not use the tests at all. Simpler was better. Make clinical decisions on the basis of growth-rate charts. If the line was not gradually sloping upward, give the child a trial of the hormone.

The pro-treatment camp was impatient with what it considered overly rigid definitions of disease. Shortness, they were convinced, was a grave disability and its victims suffered deep psychological scars. Very short children were teased relentlessly on the playground (their lunch taken and hung just out of reach) and humiliated in the classroom (always marching first in a line organized according to height). In adulthood, they faced curtailed life chances. Short men earned less than tall men; one study concluded that people who were 5 feet 6 to 7 inches tall made $2,500 a year less than those who were 6 feet to 6 feet 1 inch tall.[49] Short men also had fewer options in marriage because social conventions did not allow them to date taller women. Language itself discriminated against them, as in "getting the short end of the stick" or "coming up short." And a surprising number of medical journal articles cited as a relevant fact that since 1900, the taller of the two presidential candidates had won all but two of the elections.[50]

Did growth hormone actually increase height in normal short children? Although the answer to this question should have been determinative—who would advocate for a futile treatment?—it was not. There was good evidence that the hormone produced a spurt in growth among short children, at least during the first year of treatment. But whether the spurt resulted in a positive gain in final adult height remained uncertain. An occasional journal article reported growth hormone provided several additional inches, but typically, the number of subjects in the studies was small. Besides, other articles reported no change in final height at all following treatment.[51] The equivocal findings spurred conflicting interpretations. To one camp, the lack of demonstrable final adult height was another powerful reason not to intervene. To the other, a growth spurt in and of itself brought psychological benefits, even were final height unchanged. And some found the treatment itself and the promise of benefit (whatever the data) sufficient justification. As two University of North Carolina pediatricians insisted: "The question that should be asked is not whether GH therapy will produce a taller adult, but whether it will produce a better [adjusted] adult."[52]

Given all the unknowns, advocates for growth hormone wanted to give short children a six-month trial of the hormone and, if they had a spurt in growth, to continue with it for several years. They also raised the possibility that should subsequent research confirm the efficacy of growth hormone, then physicians who had withheld it had missed, forever, the chance to treat. Once a child entered late adolescence, the growing stage was over.[53] When it was now or never for treatment, it was prudent to treat.

The closer one examines the split, the more significant is the cure-enhance divide. The endocrinologists most concerned about the overuse of growth hormone shared a classical, really old-fashioned, definition of disease, taking laboratory data as the only reliable marker of illness.

When a physiological test revealed an abnormal score, then disease was present and cure should be sought. "Health," as one pediatrician insisted, "is not social or cultural, and is not defined in relation to others. It is a property of biologic entities."[54] Many bioethicists readily agreed. "Abandoning the treatment/enhancement distinction," observed one of them, "does not just open the door to GH therapy for short normal children. . . . It begins a cascade of changes in the scope of medicine that would forever change its face and might threaten the social consensus that gives medicine the strong moral grip it has on us."[55] Loosening the reins around disease categories would subvert society's collective ability to differentiate the normal from the abnormal, the genuine needs of patients against the frivolous desires of consumers. And the more the lines were blurred, the more impossible it was to realize a just health care delivery system and to enact national health insurance. "To expand the entitlement pie at a time when we hardly know how to divide the present pie, and at a time when many children do not get pieces of even that pie, would seem foolish."[56]

Endocrinologists prepared to dispense growth hormone focused less on societal issues and more on the well-being of the individual patient. If shortness brought penalties, the physicians should try to increase the height of the child, not alter social attitudes or worry about national health insurance. They riveted their attention on the specific child, not the larger system. They also invoked several decades of research in the social sciences and the history of medicine that fully demonstrated the ambiguous character of such concepts as disease and cure. Disease was always a socially constructed category, subject to fluctuating definitions by place and over time, with cases in point ranging from alcoholism to homosexuality (at one moment a crime, at another an illness, at still another a legitimate choice). Bringing this orientation into pediatric endocrinology meant that laboratory tests should not be allowed to dictate treatment decisions. "The point is," as one interventionist pediatrician concluded, "that short children of equal height have the same handicap regardless of cause."[57] To justify one intervention as a "cure" and denigrate the other as "enhancement" made no sense.

From this perspective, the problem was not whether gym-door fathers should be allowed to make their children into professional basketball players but how best to respond to the indignities of every-day life that very short children suffered. The stories these pediatric endocrinologists tell is of the thirteen-year-old girl who at 4 feet 6 inches has to shop in the kiddie section of the clothing store while her friends go to the teenage or adult section. Or the twelve-year-old boy who at 4 feet 9 inches is the mascot of his class or the butt of jokes. Or the twenty-one-year-old who is too short to fulfill her ambition to join the military. These physicians will offer such patients growth hormone, hoping to make them taller or, at least, happier.

V

Perhaps the most dramatic example of how tenuous the divide was between cure and enhancement appeared in 1990, when the National Institutes of Health organized a clinical trial to test the efficacy of growth hormone for short, non–hormone deficient children. Its purpose was to determine, once and for all, whether the intervention worked. The design was rigorous: Half the children would receive three injections weekly with the hormone; the other half would receive three injections with a placebo (saline solution). The children would then be followed for eight to ten years (with both groups getting between 600 and 1,100 injections), and the results would finally settle the question of efficacy.

In 1993, two advocacy groups, the Physicians Committee for Responsible Medicine and the Foundation on Economic Trends, led by Jeremy Rifkin (best known for his efforts to stop several human and plant genetic experiments), urged the NIH to abort the research. That effort failing, they sued in federal court to halt the protocol. Their arguments echoed the positions of the anti-treatment school, emphasizing that shortness was not a disease and that the hormone might not be safe. Although the court rejected their plea, the questions were troubling enough to have the NIH take the unusual step of convening a special outside panel to review the ethics of the research.

The panel, composed of endocrinologists, biostatisticians, lawyers, and bioethicists, opened their report with an estimate that some 15,000 children were receiving growth hormone, approximately half of whom were not hormone deficient. Because of the intrinsic difficulty of determining hormone levels, the panel believed that the number who were not hormone deficient was certain to increase in the coming years. Clearly, then, "determining the efficacy of hGH therapy is an important issue in pediatric endocrinology." But was it important enough to justify the protocol? Was it appropriate to give the control group hundreds of injections of a placebo over many years? The subjects, after all, were children, and federal regulations governing human experimentation considered children a "vulnerable" group, in need of special protection. Section 46.406 of the regulations contained the relevant standards. The section title was long but directly on point: "Research involving greater than minimal risk and with no prospect of direct benefit to individual subjects, but likely to yield generalizable knowledge about the subject's disorder or condition." It allowed such types of research to be conducted only if three criteria were satisfied: (1) the additional risks over the "minimal" had to be minor; (2) the intervention had to be "reasonably commensurate" with the subjects' medical or social situations; (3) "the expected knowledge is of vital importance for the understanding or amelioration of the subject's disorder or condition."

All but one of the panel members believed that the research met the first criterion—the risks were not great. The dissenter found the placebo injections too discomforting and inconvenient. The majority admitted that no one really knew what the experience of regular injections over years would mean to the children but, rather than prohibit the research, they recommended administering a survey. As for the second criterion, the research being reasonably commensurate with the children's experience, the committee found that the standard itself was highly ambiguous but that since the children would be getting comprehensive and rigorous medical evaluations, the protocol was acceptable.

The most difficult issues involved the third point, that the research had to be of "vital importance" for understanding or ameliorating a "disorder or condition." It was here that cure and enhancement faced off. The panel first argued that the standard of "vital importance" was satisfied because growth hormone was being used widely, and therefore warranted investigation in the name of public health. But what about the clause that the research had to address a "disorder"? To approve the research, the panel would have to make short stature into a disease, which it then did. It emphasized "functional impairment and psychosocial stigmatization." "Children and adults with extreme short stature may experience difficulty with physical aspects of the culture designed for individuals taller than themselves (e.g., driving a car). They may also be harmed by deeply ingrained prejudices resulting in stigmatization and impaired self-esteem." Did these disadvantages make shortness into a disease? Apparently yes, for the panel approved the protocol.[58]

The fundamental lesson to be drawn from the NIH experience was that if a condition causes unhappiness, psychological pain, and social disadvantage, then it represents a disease, and interventions to remedy it should be considered cures. Indeed, the NIH attitude affected the FDA. An advisory panel on the use of growth hormone for short, non-hormone deficient children met in June 2003 and recommended that very step. Some of the panel's members were unhappy with the decision, complaining about the medicalization of shortness and the absence of data about long-term treatment effects. But the majority supported the intervention on the grounds that some children did experience a spurt in growth and that the stigma of shortness and the psychosocial pains associated with it were severe.

Thus, any and all attempts through endocrinology now, or genetics in the future, to try to reduce unhappiness, pain, and disadvantage, whether the result of physical appearance, memory capacity, sleep patterns, muscular strength, or advanced age, have legitimacy. Truly, cure and enhancement are becoming one.[59] In July 2003, the FDA accepted the recommendation, approving HGH for normal but unusually short children.

VI

No one had a greater stake in eliminating the distinction between cure and enhancement than the pharmaceutical companies producing growth hormone. A market composed exclusively of children with laboratory-defined hormonal deficiencies would amount to no more than 7,500 children a year, a small enough number to qualify the drug for "orphan" status, so little prescribed as to warrant special federal incentives for its continued production. But were the market to include short children, then the numbers could swell to 100,000, perhaps 200,000—and given potential profits, companies would happily forgo the orphan drug designation. It became the mission of Genentech to have its drug, Protropin, reach that larger market well before the FDA decision.

To this end, Genentech's strategy, as pieced together in congressional hearings and a federal criminal trial in Minnesota, relied on all the standard promotional approaches, including medical journal advertising, free samples to doctors, and visits from detail men, and now, detail women. But it also invented new strategies and embellished old ones, in the process demonstrating the exceptional difficulty of regulating the conduct of pharmaceutical companies and their physician allies.

One of Genentech's first forays was to help underwrite the costs of a foundation whose mission was to alert parents and school staffs to growth disorders. The Human Growth Foundation (HGF) was founded in 1965 by a small group of concerned parents, assisted by Robert Blizzard, the first head of the National Pituitary Foundation. It attracted some one thousand families as members and had forty-two chapters across the nation. The HGF served as a support group for parents, as an advocacy group, as an educational resource, and as a grant maker to encourage young researchers to enter the field. After 1985, Genentech became its single biggest funder, accounting for one-quarter of its budget, and several of the company's executives sat on the foundation's board.[60]

One of HGF's major activities, paid for by Genentech, was to fund elementary and secondary school screening programs. It provided materials and personnel to train school staff on how to measure student growth and maintain growth charts. Parents of children who fell below the 5th percentile for their age group received a letter advising them of this fact and recommending that they see their doctor. The letter contained no manufacturer's name and did not recommend specific physicians or treatments.

When the foundation and Genentech officials were asked by a congressional committee about the propriety of this joint venture, their response was quick and confident. This was a health program, alerting parents to potential problems that might be hormonal in origin, or nutritional or physiological (including the possibility of pituitary tumors). "Growth screening is one component of essential medical preventive care," the foundation noted. To screen, educate, and facilitate referrals were unrelated to any "alleged marketing efforts of any pharmaceutical company."[61] Genentech added that seven states had mandated such screening programs on their own, thereby demonstrating their value. "If the focus of the program is public health and if it supplants largely unavailable public funds, it is a public good."[62]

Others, however, found the program entirely too self-serving. Because Genentech had a financial stake in increasing the number of short normal children who used growth hormone, the screening effort seemed obviously designed to expand its customer base. In one sense, the charge was a stretch. No one had ever accused school dental, eyesight, or hearing screening programs of being fronts for physicians or dentists, or charged that a school lice inspection program was a get-rich scheme for dermatologists and shampoo manufacturers. But height screening seemed different, perhaps because Genentech had a well-earned reputation for aggressive marketing or because shortness was not the equivalent of cavities or lice. In any event, criticism mounted to the point that Genentech ended its financial support.

The HGF and Genentech experience made two points eminently clear. First, from a marketing perspective, the tactic made sense and worked well. Once a disease category was created, particularly in a shadowy and new area, then the more cases diagnosed the greater the company sales. Second, from a regulatory perspective, it was nearly impossible to distinguish the fair from the foul. Health education and disease prevention were legitimate activities—and no one could delineate where health prevention left off and crass self-interest began.

However advantageous it was to work with parents, schools, and the communities, Genentech devoted greater time, energy, and money to physicians. Here, too, the propriety of its actions evoked criticism and, in one case, a substantial fine. But again, it had ample reason to insist that it was only doing what everyone else did. The major company activity involved post-marketing surveillance of growth hormone treatment. Since the FDA had moved quickly to approve its drug and the clinical data available from the pituitary hormone period was, of necessity, limited, it seemed appropriate to continue to collect information on the efficacy of the drug as it was used in doctors' practices. Genentech was eager to underwrite the costs of the effort, not so much because of the information itself—since the drug was already approved, new data might hurt as well as help its marketing—but because of the ties it could forge with the prescribing physicians.

The arrangement designed by Genentech gave it a seemingly legitimate way to persuade doctors to prescribe its product and then reward them financially for doing so. Some would label it a kickback; the company insisted it was a research grant. And often, albeit not always, it was impossible to say who had the stronger case.

In order to carry out post-marketing surveillance of growth hormone, physicians had to maintain careful records of the patients they treated, the dose level, the duration, the side effects, and the outcome. Each child had to have his or her own chart, scrupulously maintained and updated; the entries were to be reported to a central office. It was a time-consuming procedure, although not substantially more than any meticulous pediatrician might do. Genentech's contribution, in neutral language, was to reimburse physicians for keeping the records and then cover the costs of their travel to meetings to share their findings with colleagues. As with its support of the Human Growth Foundation, the company's activities could seem to be not only legitimate but very valuable: imagine what could be learned from the data and taught to physicians. But in more skeptical terms, post-marketing studies gave Genentech the opportunity to pay doctors to prescribe growth hormone. The company's payments to physicians varied, but several thousand dollars per child treated was not uncommon. Needless to say, Genentech chose the best hotels and resorts for their meetings—always warm and sunny—with spouses invited and paid for. Genentech's detail men often filled out the forms and did the paperwork for the doctors, facilitating contact with them. In short, post-marketing surveillance was an ideal umbrella under which to distribute company largesse and promote its product.

In at least one instance, the bounds of legality were overstepped. In 1994, the U.S. attorney general's office indicted Genentech's top sales executive and officials at Caremark, a home health care provider with exclusive rights to distribute Genentech's drug outside of hospitals, for paying kickbacks to a Minnesota physician, David Brown. Over an eight-year-period, he reputedly received $1.1 million in return for prescribing Protropin. Caremark settled its part of the case with an admission of guilt and a payment of a $110 million fine. Brown was also indicted, pleaded innocent, and came to trial in federal court in August 1995. The outcome was more uncertain than might be anticipated in light of Caremark's admission of guilt because the judge ruled at the outset that Caremark's settlement was inadmissable as evidence. To let the jury learn of it would be too prejudicial to the defendant.[63]

Over the course of the trial, no one denied that Genentech had paid $1.1 million to Brown. Indeed, Brown had recorded every dollar he received. The open question was whether the money represented bribes and payoffs for the prescriptions he had written or reimbursement for legitimate research and consulting activities. Brown's defense was simple. Part of the money went to cover the costs of his research. After all, Genentech, like other pharmaceutical companies, dispensed research grants and he, like thousands of other investigators, had accepted one. He had conducted the investigations, as evidenced by the research abstracts he had presented at annual pediatric endocrinology meetings. No matter that he had never published an article in a peer-reviewed journal or, for that matter, in any journal.

Brown insisted that Genentech's other payments to him, which included a percentage of his nurse's salary, were to facilitate his participation in its post-marketing survey. Data on his 200 to 300 patients were incorporated in the major Genentech study of growth hormone results and in every one of its smaller subgroup studies. Even the fact that his total payments were far greater than those received by other physicians could be rebutted. The company did not set a fixed or flat fee for services rendered. As a Genentech official testified on his behalf: "I mean for any study, for any doctor, there is a negotiation, if you will, because costs differ in different places. You can't develop one budget and try to implement that throughout the United States because things cost more in New York than they do in Iowa." To be sure, Genentech conceded that it never audited Brown, or for that matter any other physician. "We don't audit, you know, individual people for the time they spend on the work." The company trusted the doctors.

The prosecution, for its part, derided the quality of the research that Dr. Brown performed, stressing that it was never published. More telling, the funds paid to him came not from Genentech's research division but from its marketing and sales division. The prosecution also read from documents showing that Brown negotiated a fee that gave him 5 percent of all proceeds from his prescriptions of growth hormone and that Brown was the highest dispenser of Genentech's Protropin in the Minnesota area. To clinch its case, the prosecution brought in Dr. Robert Ulstrom as an expert witness to testify that Dr. Brown inappropriately prescribed growth hormone to his patients, seemingly practicing bad medicine in return for good money. Ulstrom, a retired endocrinologist from the University of Minnesota, had done substantial research over the period 1954 to 1964, but had published only two papers on growth hormone. Why did the prosecution select a physician who was retired and not particularly expert in growth hormone? Because all the other experts that the government wanted to use had ties, loose or otherwise, to Genentech or to other companies producing growth hormone.

The prosecution led Dr. Ulstrom through seven of Dr. Brown's patient records, trying to establish a pattern of overuse of growth hormone. The first case was Mark Miller, a thirteen-year-old who, when he first saw Dr. Brown, was only as tall as the average ten-and-a-half-year-old. His father was short, so was his sister, but his mother was 5 feet 8 inches. Reviewing the case, Dr. Ulstrom opined that Miller was not hormone deficient as measured by challenge tests and overnight blood tests but, nevertheless, Dr. Brown had prescribed growth hormone. Miller did enjoy a spurt in growth, which Dr. Ulstrom explained was because the boy was entering puberty and the drug had a short-term but not long-term effect. And so it went with the other six cases—children who were not growth hormone deficient but to whom Dr. Brown had given the drug anyway.

The defense had an easy time demolishing Ulstrom's testimony, not only because his credentials were weak but because it had no difficulty finding professional support for Dr. Brown's decisions. First, it stressed how arbitrary the definition of growth hormone deficiency was. "One of the things that kind of plagues this area," testified a Genentech official, "is that there is no single object measure that gives a clear answer as to a specific child who should or shouldn't be treated with growth hormone. . . . And so this is an area where particularly physicians' judgment is important."[64] Second, physicians were right to base treatment decisions on more than laboratory findings. Other considerations "weigh heavily on their mind—as to how psychologically affected the child is, how many problems they are having in school, how many problems they are having with their peer group."[65]

As for the specific cases, the defense obtained a letter from one of Ulstrom's Minnesota colleagues justifying Brown's prescriptions. "Although we do not usually treat children with adequate responses to growth hormone stimulation studies . . . there is suggestive evidence that such treatment may be indicated in 'normal' short children." It also introduced data from a recent survey indicating that 82 percent of pediatric endocrinologists had on occasion given growth hormone to children who were not by standard laboratory definitions hormone deficient. The defense also brought in a more up-to-date endocrinologist, Ron Rosenfeld of the University of Chicago, who testified that Brown's treatment decisions were well within the professional standards in the field. In fact, on cross-examination, Ulstrom himself conceded that laboratory tests had a number of "limitations," and findings of hormone deficiency were more in a "gray zone" than definitive.

Because Mark Miller had grown taller, a courtroom dialogue ensued that revealed how different judicial standards of evidence and proof were from scientific evidence and proof. To Dr. Ulstrom, the fact that Miller had come within two inches of his predicted adult height was irrelevant. As a physician and researcher, the null hypothesis ruled—that is, Miller might well have achieved that height anyway. The burden of proof was on Dr. Brown to establish that his treatment had been the determining factor. But the courtroom adhered to a different standard. The presiding judge interrupted Ulstrom to explain that a criminal charge had been levied, and the burden of proof in a criminal case rested with the prosecution, not the defense. It was not Dr. Brown who had to demonstrate efficacy of treatment but Dr. Ulstrom who had to refute it. Unless the prosecution could prove beyond a reasonable doubt that growth hormone was not responsible for the growth, Dr. Brown was innocent of the charges.

The end of the case brought several surprises. Despite the able defense, the jury convicted Brown of taking kickbacks. But the judge then set aside the verdict because of juror misconduct; one of them had learned about Caremark's admission of guilt and shared the information with his fellow jurors, who then discussed it in the course of their deliberations. The judge ordered a new trial, but the federal prosecutors decided not to proceed. This prompted several parents whose children had been treated by Dr. Brown to file their own civil suit against him for failure to disclose the kickback scheme; they could not pursue a malpractice claim because the statute of limitations had run out. The court, however, refused to hear their case, on the grounds that it was a malpractice suit in disguise, not a real case of consumer fraud.

When all was said and done, Dr. Brown walked away with no criminal penalty and no fines, in the process providing an object lesson.[66] Criminal prosecution is rarely an effective means for regulating medical practice. And it certainly has no chance of succeeding when the prosecution must rest its case on a distinction between cure as good medicine and enhancement as bad.

VII

The growth hormone story does not end with short children—it has a second life with healthy older men. The logic for giving them the drug has a familiar ring. The fact that growth hormone levels in the elderly are lower than in the young suggests the possibility that a hormone deficiency is the cause for their frailty. Since HGH improved the physical condition of children, it should improve the physical condition of the elderly. Grandparent as well as grandchild might both reach for their growth hormone vials.

The first step was giving growth hormone to patients whose pituitary glands had been removed or radiated to prevent the spread of tumors.

In 1962, writing in the *NEJM*, M. S. Raben reported that he treated a thirty-five-year-old schoolteacher suffering from pituitary gland insufficiency with growth hormone three times a week; after two months, the patient experienced exceptional physical and psychological benefits, including "increased vigor, ambition, and a sense of well-being."[67] The case did not immediately spur further research or change clinical practice because of the acute pre-1985 shortage of growth hormone. But once Genentech synthesized the hormone, physicians began to administer it to adults with impaired pituitary functioning. One British team reported that growth hormone given to twenty-four adults with severe growth hormone deficiencies reduced body fat and cholesterol and increased muscle mass. Although some patients experienced fluid accumulation and joint pains, the benefits outweighed the risks.[68]

The findings were well received because they were consistent with the effects of growth hormone in children. Since the drug increased muscle strength and reduced body fat among hormone deficient youngsters, it should exert the same effects in hormone deficient adults. Moreover, the fifteen-year experience with giving synthetic growth hormone to children had not produced serious side effects, which made it seem all the more reasonable to administer it to pituitary-damaged adults.

In no time at all, the distinction between pituitary-damaged adults and normal adults evaporated. If growth hormone improved body composition in one cohort, why not try it for another? Skeptics noted that an age-related increase in fat might be the cause, not the result, of a decrease in growth hormone production, so interventions should emphasize weight reduction, not the administration of a drug. They were also dubious about generalizing from children to adults, and from older patients with pituitary disease to older patients who were healthy—perhaps only those with severe deficits benefited. Nor were they convinced that older men and women were actually growth hormone deficient. Why set the normal level by the teenager and consider all sixty-five-year-olds deficient? Why not age-adjust the measure, taking the average sixty-five-year-old reading as normal? In effect, the debate about levels was a debate about the wisdom of nature. Were the lower levels of growth hormone protective or injurious? Did we tamper with nature at our peril or to our benefit?

Despite these larger questions, research on HGH in adults continued. In 1990, Daniel Rudman at the University of Wisconsin reported the first results of administering growth hormone to a group of healthy men between the ages of sixty-one and eighty-one. He gave twelve of them injections of growth hormone three times a week, and gave a control group of nine men a placebo. At the end of six months, the growth hormone recipients demonstrated an average increase in body muscle of 9 percent, a decrease in fat tissue of 14 percent, and an increase in skin thickness of 7 percent—findings that Rudman presented as a "reversal of the effects of 10 to 20 years of aging." (This phrase, as we shall see, had a very special appeal to the media and drug companies.) The control group showed no significant changes at all. Side effects appeared very minor; none of the recipients had edema or increases in sugar levels or blood pressure. Rudman's conclusion was that reduced levels of growth hormone in otherwise healthy men were responsible, at least in part, for the loss of muscle, gain in fat, and thinning of the skin, and these changes could be reversed by administering growth hormone.[69]

These claims, as would be expected, spurred further research. As might also be expected, there was no consistency in the design of the studies. Teams used different dosages of growth hormone, different measurements of growth hormone levels, different populations, different lengths of administration, and different body measurements. There was also no consistency to the findings. One Stanford group reported much less favorable outcomes and a dropout rate of 11 of the 19 subjects because of the severity of side effects, including carpal tunnel syndrome and severe joint pain. Other teams found that growth hormone strengthened the lumbar spine but not the hips. Still others detected no effects whatsoever on adult bones: "It is difficult to justify optimism that any tolerable dose of GH will provide a major skeletal anabolic effect in elderly men and women."[70] One team gave growth hormone to healthy sixty-four- to seventy-six-year-olds every evening for six weeks and learned that muscle strength increased significantly as measured by the use of a rowing machine, increased only slightly as measured by a shoulder press or leg curl, and increased not at all by other strength tests. Body weight, body fat, body muscle, and cholesterol levels were all unchanged.[71] Another team reported "no evidence of an association between GH secretion and muscle mass," or between short-term hormone administration and bone mineral density at the forearm, spine, or thigh.[72]

The negative evidence continued to mount over the 1990s. One particularly thorough study by Maxine Papadakis at the San Francisco Veterans Administration Center, published in the *Annals of Internal Medicine*, involved the effects of growth hormone in fifty-two healthy adults of an average age of seventy-five. Half the group received the hormone for a six-month period, the other half received a placebo. Those receiving the hormone did have increased body muscle and decreased fat, but the changes had no impact on functional ability. Both groups performed the same on tests of muscle strength and endurance. Those with the hormone did better on some cognitive tests and worse on others. There were no differences in mood or in measures of depression. However, there were major differences in side effects. Recipients commonly reported edema and joint pain. In all, physical exercise (which cost nothing) led to better performance levels than growth hormone (at $18,000 a year), and with no side effects. Papadakis's conclusion: "Growth hormone should not be used to preserve or improve function ability in healthy, functionally intact older men."[73]

In light of these negative findings, several teams took another look at the use of growth hormone in patients with pituitary tumors and discovered many more serious side effects than originally reported. It turned out that adults with growth hormone deficiencies responded differently than children; they were far more likely to suffer major adverse events, including edema, numbness, and joint pain. The worst side effects were experienced by those who had been considered most likely to benefit from the hormone, the overweight patients.[74] So too, the benefits turned out to be minimal. Although some patients reported genuine gains, the great majority did not. "A definite improvement in well-being when replacement GH is given to large groups of patients have [*sic*] yet to be established."[75]

Thus, by the end of the decade, the consensus among investigators was that growth hormone served no purpose in otherwise healthy, older men. It altered body composition but offered no "significant improve-

ment in muscle strength or exercise tolerance."[76] At a minimum, as
one reviewer commented in 1999, the case for it was "certainly 'not
proven.' "[77] That same year, the National Institute on Aging of the NIH
issued a guideline against giving growth hormone to healthy adult men
because "too little was known about it." It went on to state that neither
growth hormone nor anything else should be given "as an anti-aging
remedy, because no supplement has been proven to serve this purpose."[78] Nevertheless, a vocal minority did not want to abandon growth
hormone.[79] As the members of a prestigious growth hormone research
society insisted: "It is likely that GH replacement will in the near future
become as routine . . . as sex hormone replacement."[80]

The irony of this prediction aside, growth hormone use soon became
even more problematic. A few small studies had found that growth hormone helped patients with severe illness to recover more quickly, particularly patients in intensive care units who had undergone cardiac or
abdominal surgery. The findings were of such potential therapeutic
importance that they stimulated a multi-center research project by a
consortium of Finnish and Western European investigators. The teams
gave 250 patients in intensive care units high doses of growth hormone
and gave another 250 a placebo. The results, published in 1999 in the
NEJM, were as unambiguous as they were unsettling. The group
receiving growth hormone was far more likely to die in the hospital than
the group on the placebo (roughly 40 percent compared to 20 percent).
Even among survivors, those treated with growth hormone had worse
exercise tolerance than those on the placebo. Why the outcomes should
have been so negative was unclear; the speculation was that growth hormone impaired the body's immune defenses. But whatever the reason,
as the editorial that accompanied the publication of the finding declared:
For now "growth hormone should not be given to patients with critical
illness."[81]

Another shadow fell on growth hormone a few months later when
researchers examining physical growth and life span in a special breed
of mice—who had a gene knocked out so as to decrease their levels of
growth hormone—reported a surprising finding: the mice who were
growth hormone deficient lived considerably longer than mice with
normal levels of growth hormone. "The results," they wrote, "impli-

cate GH deficiency as the major factor in increased longevity and suggest the use of a cautionary approach to the therapeutic administration of GH, especially as an anti-aging agent, until more studies can be completed." It was possible that growth hormone was not the culprit; the gene that had been knocked out may have performed other vital biological functions, so that its absence produced a general debility. But here was another piece of evidence that higher levels of growth hormone might be life-shortening. The verdict of "not proven" was changing to "dangerous to your health."[82]

In retrospect, healthy older consumers would have been wise to avoid the risks of taking growth hormone. But how were they to know that caution was appropriate? Not from reading the initial press accounts. Although some reports of Rudman's 1990 University of Wisconsin paper were circumspect,[83] the Associated Press release had as its opening line: "Hormone injections can reverse some of the damage of aging and give people back the firmer flesh of their younger years." It went on to explain that growth hormone, now available in greater supply, would "help elderly people build up sagging muscles, take off flab and grow more youthful looking skin—turning back the clock as much as 20 years in just six months." Moreover: "The volunteers who got the shots said treatment made them look better and feel stronger, and their wives agreed."[84] The AP was not selling growth hormone, it was selling stories, and the hook lay in the effectiveness of growth hormone, not the preliminary nature of the findings. Its account had to be positive— and so the promise that growth hormone would reduce flab and make you feel stronger, and heighten sexual performance. Although in interviews Rudman himself sounded notes of caution, he often repeated the "twenty years younger" line, which in its very specificity gave credence to the intervention and aroused still more interest.

Endocrinologists soon were telling reporters that they were deluged with requests for growth hormone, which only served to increase interest. "They've called me at home—people wanting growth hormone for their parents," commented one expert, Mary Lee Vance. Some physicians worried about a black market in the hormone, and rumors were circulating about offshore clinics dispensing growth hormone. But the "avalanche of demands for immediate access to the hormone" delighted

Rudman. Whatever else, the frail elderly would no longer be "written off as hopelessly debilitated."[85]

Or so it seemed until the mid-1990s, when the doubts intruded. The press occasionally reported on the expanding list of side effects and some physicians' skepticism—the elderly on growth hormone "don't suddenly want to go cartwheeling down the corridor." Then came the Papadakis article, and although the *Annals* is not a journal as thoroughly covered as the *NEJM,* it is visible enough for a story on negative findings to create a stir. In this instance, the press did not bury the retraction. *USA Today* ran a front-page headline: "Elderly See Few Benefits from Growth Hormone." To be sure, the *New York Times,* which had put Rudman on page 1, put Papadakis on page 13, but its account did open with Papadakis's remark that "We cannot recommend it. . . . It's not the fountain of youth." And in a corny but not inaccurate rendition of the findings, the banner at *New York Newsday* ran: "Fountain-of-Youth Springs a Leak."[86]

Thereafter, a far more skeptical tone dominated media accounts of growth hormone. Individual endorsements ("I feel like I have restored my body to what it was like in college") were counterbalanced by physicians' warnings ("I tell everyone who calls, 'Don't take it!' "). When the scary report on growth hormone and mortality among ICU patients appeared and was soon followed by the finding of increased life span among growth hormone deficient mice, the press reported both stories as reinforcing already existing doubts on growth hormone efficacy.[87]

One might have imagined that all this would bring closure to the growth hormone story—a roller-coaster ride, to be sure, but everyone getting off safely, explosion followed by implosion, and final verdicts reached in relatively efficient fashion. But that is not the final word. Growth hormone use is ongoing, providing an object lesson in the challenges that Americans confront and will continue to confront in evaluating and regulating would-be enhancements. First, HGH promotion goes out through the modern version of traveling snake oil salesmen, reinvented as Web "health" sites. Second and even more distressing, a number of physicians, clinics, and societies unabashedly promote it, along with other purported enhancing techniques. These marketing ploys have escaped both regulatory authority and professional discipline.

The snake oil first: At least a dozen Web sites advertise and sell substances purported to be growth hormone. Thus, hgh-pro.com offers "the 'proven' most effective Human Growth Hormone product available without a prescription." It is dispensed by a nurse, "a medical professional actually involved with the clients, rather than a salesperson with no medical training." Buy two bottles at regular price and get a third free. Results are guaranteed: improved energy, muscle mass, hair growth, immune system, mood, sexual performance, and a sense of well-being. Exactly what the product is is never made clear. It cannot be growth hormone itself, since that requires a prescription. Rather, these are supposedly natural substances, whatever that may mean, with the promise that they are the market's best.

Go to the World Health Network to have a choice of purveyors of growth hormone–like substances. There is A Physician's Blend HGH or Regenesis Plus (an "elite growth hormone product" in an oral spray). Regenesis is also available at young4ever.com. In the first month, it will bring more vivid dreams and better sleep along with heightened energy; in the second month, it delivers weight loss and enhanced sexual function. By the sixth month, it will lower blood pressure and improve cholesterol levels. AgeForce sells it too, urging consumers to buy the product from them since they are a pharmaceutical manufacturer, not a marketing company. AgeForce also provides a bibliography on growth hormone, with selective quotations from the literature, including Rudman's statement on reversing twenty years of aging. Needless to say, Papadakis is not on the reading list and no mention is made of the recent ICU findings or the results with mice. Lifespanlongevity.com promises to effect "a dramatic slowing and even reversal of the aging promise" and backs up its claim by noting that the *NEJM* has reported that human growth hormone can reverse "the aging process . . . from TEN to TWENTY years." Of course, the column on "What researchers have to say about hGH" leads off with Rudman. All the Web sites close with the same tag line: "The information provided has not been evaluated by the FDA and is for educational purposes only. It is not intended as a diagnosis, treatment, for prescription for any disease. Consult your physician."

Seeing a doctor, however, does not provide as much consumer protection as might be hoped. A growing number of physicians present themselves as "anti-aging" specialists and their calling card is access to hormones, often growth hormone. In the early 1990s, a dozen physicians created the American Academy of Anti-Aging Medicine (A4M), which now claims 8,000 members. It administers an examination that gives physicians "board certification" in anti-aging medicine. It sounds impressive—after all, board certification is the gold standard for specialties—but it is entirely fictive, not recognized by the Council on Graduate Medical Education, which governs specialty certification. The unwary or unsophisticated consumer, not likely to know this, might believe that A4M is a genuine specialty board, akin to those in internal medicine or surgery.[88]

Members of the A4M are as aggressive as they are unembarrassed by the problematical nature of their practice. They boast, in the tradition of Brown-Séquard, of being their own best customers, perfect advertisements for the wonder of growth hormones. Dr. Ronald Klatz, president of A4M and author of *Grow Young with HGH,* claims that since taking the hormone, his waist lost two inches and his chest expanded two inches; he is chronologically forty-four years old, but only thirty-four biologically.[89] Dr. Alan Mintz, head of the Cenegenics Anti-Aging Center, declares: "I've been taking hGH for 3 years, and I've never felt better or stronger."[90] Dr. Adrienne Denese, who practices anti-aging medicine on Manhattan's wealthy East Side, injects herself as well as her patients with growth hormone. "It takes a few days to feel it mentally, and the physical effects—like extra muscles—you can see in a month."[91] In the tradition of Gertrude Atherton, physician testimonials go hand in hand with celebrity testimonials, including those of Oliver Stone and Nick Nolte.

Estimates of how many Americans take growth hormone for enhancing physical capacity range from 5,000 to 30,000 to 250,000. The annual report of Pharmacia, which sells Genotropin, now the best-selling growth hormone product in the United States, says its sales are up 25 percent to a total of over $475 million. But how many of the purchasers are children, adults with pituitary disease, athletes, or otherwise healthy adults remains unknown. One Web site, humangrowthhormonesales.

com, sells Lilly's growth hormone product. (Purchasers must have a doctor's prescription, which the company will honor for one year.) Between February 24, 2001, and June 14, 2001 (when we logged in), the site had 7,686 visitors. But log-in does not mean purchase, as witness our own visits.

At least 100 clinics advertise themselves as anti-aging medicine centers. Most of them, as might be guessed, are located in California and Florida, with ample numbers in New York, Nevada, Arizona, and Texas. The patient base is not enormous—most report 50 to 150 clients—but Las Vegas–based Cenegenics claims 1,200. (Note the self-promoting name, *cene* to suggest centenarians, *genics* to suggest, incorrectly, that genetics is involved.)[92] The fees are hefty and for the most part not reimbursed by insurance companies; they range from $1,750 for a workup at Cenegenics to almost $5,000 at Lifespan, with follow-up packages of hormones, vitamins, and enzymes adding another $1,700.[93] Adopting the model of personal trainers, some clinics, like the Anti Aging Medical Associates of Manhattan, set an annual fee, with patients entitled to unlimited visits. Only rarely will a clinic advertise "cost effective" services with "discounts available."

Some clinics are run by plastic surgeons or dermatologists who, as we have seen, are eager to add anti-aging to their practices. But most clinics are free-standing and devoted exclusively to anti-aging. What that means, first, is a vast and truly bewildering array of diagnostic tests. One such clinic runs patients through a roster of laboratory tests that takes 22 pages to describe, so as "to assess more than 150 relevant biomarkers of aging," as though anyone knew what the markers were. The workup includes standard blood tests and examinations (like an EKG), but also (with a disclaimer "for investigational purposes only") blood or urine levels of beta carotene, hydroperoxides, tocopherol, melatonin, and trace metals (tin, zinc, mercury, nickel), oxidative stress measures, and, of course, hormone levels. Once the tests are completed, it moves to therapeutics. Through the administration of growth hormone, it promises to restore "youthful levels," which will produce more muscle, reduce fat, improve bone density, improve short-term memory, and enhance "clarity of thought." Another competitor, the Institute of Anti-Aging Medicine, located fifteen minutes from Houston's airport, claims that "through the use of state-of-the art technologies such as comprehensive hormone replacement therapies, including growth hormone, the aging process is dramatically retarded, even reversed." The California Anti-Aging Institute in San Diego, headed by Dr. Ron Rothenberg, who is board-certified in (of all things) emergency medicine, provides "hormone replacement programs for men and women including human growth hormone." The promised results: "the reversal of the aging process and peak human performance."[94]

Although anti-aging physicians will sometimes acknowledge that their interventions do carry risks of side effects, they are hardly concerned about them. Some keep the dose levels low, probably too low to do much good were these interventions effective, but low enough also to prevent the most harmful consequences. They are quick to discount the specter of cancer. As Ronald Klatz commented to one reporter: "If there was a risk of cancer, where are the bodies?"[95]

The clinics have not been curbed and the practitioners not reined in. The FDA is helpless to do anything about the off-label use of drugs already on the market for therapeutic purposes. As Murray Susser, who practices anti-aging medicine in Los Angeles, notes: "It's legal, and people have a choice. It's only misuse if I lie to them. I say to people who are taking it, 'It's experimental, it may help, but I don't know for sure.' " Nor can the FDA now do anything about so-called natural substances, which congressional action has removed from its purview. Professional medical societies are notably silent. The organization representing clinical endocrinologists, for example, has declared that no evidence exists for the efficacy of any of these interventions. But it does nothing more than issue such statements, perhaps taking comfort in the fact that most anti-aging practitioners are not board-certified endocrinologists. Anti-aging physicians could be sued for malpractice and it would not be difficult to demonstrate that their practice patterns deviate seriously from professional standards. But patients have not been inclined to take them to court, either because they do not want to admit publicly to seeking anti-aging medicine or, more likely, because they have found the physician sympathetic to their problems and ready to give them ample time. Or perhaps these patients are the quintessential risk-takers, determined to be the first in line, ready to suffer disappointments but come back again for the next regimen. As one frequenter of these clinics boasted: "I wasn't worried because I'd read a lot about it. But as with anything new, people are afraid."[96] All of which helps explain why enhancement technologies, whatever their putative benefits or demonstrated risks, will have significant space in our future.

EIGHT

Peak Performance

OVER THE PAST DECADE, an explosion of knowledge about the human genome has dramatically increased the prospects for refashioning the self. Although specific interventions to enhance performance are in drawing board stage, the possibilities inherent in the new technologies have sparked extraordinary public and professional interest. A spate of popular and academic books explores some of the promises of modifying genes so as to enable ordinary men and women to perform exceptional physical and mental feats. They then go on to ask what it will mean when genetic engineering is able to make us stronger, smarter, less in need of sleep, more able to run, lift, and swim. What if gene manipulation could make us more assertive or considerate or cautious or venturesome? What if genetic knowledge could double the average life span, enabling people to live far longer productive and active lives?

As the historical record we have traced makes apparent, these issues have both a personal and social dimension. As individuals, each of us now stands in roughly the same relationship to medical knowledge and practice as our predecessors did in the 1920s and 1930s. Then we were our hormones; now we are our genes. Although the quality of scientific research may have advanced, so has its sense of adventure. It is far from clear whether investigators' impressive credentials or physicians' enthusiasm will serve as adequate guides to the emerging technologies. There are already numerous examples of interventions that promised to enhance physical strength and longevity but, in fact, reduced them. Nevertheless, the benefits of being Forever Young or Forever Awake are likely to be presented with such exuberance as to obfuscate the profound risks of harm.

I

The advances in genetics are reordering our conceptions of the respective roles of environment and genetics. For most of the twentieth century, genetics was understood as a fixed and limiting factor for any single individual. In the popular imagination and in most scientific circles as well, inheritance was a lottery that produced winners and losers, determining physical features (eye color, skin color, height) and mental and behavioral traits (from the degree of memory to the type of personality). By contrast, the liberating factors emerged from the social environment. Through the right influence of family, school, and community, through the impact of good housing, diet, and health care, individuals would be able to shape their own futures.

As we know all too well, a few societies have tried to apply a program of positive and negative eugenics to improve the nation's stock. No country more brutally pursued these tactics than Nazi Germany. It encouraged some sectors of the population to reproduce (Aryans), and discouraged or prevented others from doing so (including Jews, the mentally ill or retarded, tuberculosis patients, homosexuals, and Gypsies), either through forced sterilization or murder. In less horrendous but still coercive fashion, the United States in the 1910s and 1920s incarcerated and sterilized so-called mental defectives to try to change the composition of the population. As late as the 1980s, China was using compulsory sterilization to these same ends.

However inhumane and cruel these efforts, they were futile. Because of a basic ignorance of the rules of inheritance (Down's syndrome, for example, is not inherited), or because the imposition of such controls was beyond the capacity of the state (in the United States, courts prevented some of the worst abuses), eugenic efforts came to naught. And even at their most grotesque, the programs did not, and could not, seek to alter the inherited characteristics of a given individual. The hope was that in some distant future the nation would become all white or blond or Aryan, but in the here and now people still had to play the genetic codes they had been dealt.

By contrast, liberal and reform-minded social activists designed and implemented public programs for the explicit purpose of countering and overcoming genetic influences. Governmental interventions to ameliorate the social environment were intended to expand individual life choices and in this way benefit the larger community. The Progressive agenda in the opening decades of the twentieth century epitomized the movement. Improve housing, and a constitutional proclivity to a disease like tuberculosis would be offset. Establish schools, and mental capacities would increase. Build playgrounds and organize settlement houses so that children would grow up to be cooperative and lawful. The work of the anthropologist Franz Boas typified the approach. Boas was committed to the principle that seemingly inherited traits reflected environmental conditions, and his case in point was height. The prevailing belief was that immigrant children, because of their inferior stock, were shorter than native-born children—and size was an indicator of general physical and mental capacity. Boas demonstrated that immigrant children raised in healthy environments grew as tall as native-born children. Social conditions mattered most, not hereditary traits.

Now, in a genome era, the older divide between a liberal–social environment camp on the one hand and a conservative–genetics camp on the other has weakened, almost disappeared. Take the dictum "You are your genes." Rather than encourage passivity and resignation before fate, the postulate is turning into a call for intervention. Genes can be manipulated. The organism is fluid. As soon as the biological mechanisms of heredity are understood, the genetic sentence can be altered, whether it involves cure (as in the case of an inherited propensity to breast cancer) or enhancement (as in an inherited propensity to forgetfulness). In effect, the idea of genetics as a limit is giving way to genetics as a tool for redesigning the individual.

The importance of this shift cannot be exaggerated. It means, first, that the goals of genetics are far more focused on altering individual characteristics rather than group characteristics. We are not back in the Nazi era with a dream of collective change, but on the brink of an era that promises individual change. Second, genetic engineering in the future is not likely to be imposed by the state from above, but pursued by individuals, embraced from below. The goal is to improve each one of us, not all of us through collective means. Third, the social environment is being defined as more restrictive than liberating. The quicker and easier route to change may be through genetics because manipulating a gene may be far simpler than altering societal institutions. For all these reasons, opponents of genetic engineering will not be able to invoke the experience of eugenics. The many forces driving enhancement that we have been tracing, from science to medicine to culture and commerce, will not be derailed by an appeal to historical precedents. Technologies that we once feared as coercive are being viewed as liberating.

In more specific terms, genetic research is exploring the operation of the brain, including the mechanisms of memory as well as the causes of Alzheimer's disease and schizophrenia. It is also examining the biological sources of behavior, ranging from alcoholism to temperament (introverted or extroverted, conservative or risk-taking). So too, investigators are exploring the genetic basis of longevity, looking well beyond the possibility of extending the life span by 20 percent (which occurred over the course of the twentieth century) to 50 percent, so that an average life span would be in the range of 140 to 160 years. There are formidable barriers to realizing all these agendas, but the field may be on the brink of revolutionary changes.

There is no holding back the enterprise. Not only are the goals highly attractive to many people, but even more important, the research itself begins typically with an attempt to cure a disease and then at a subsequent stage explores the possibilities of enhancement. Thus, investigations that initially seek to repair shattered memories may eventually provide clues to expanding normal memory. Research that seeks to cure mental illness may uncover mechanisms to promote optimal behavior.

Because the possibility of such accomplishments takes the breath away, the need to understand the methods of the research and their likely outcomes is essential. As this history of enhancement has demonstrated time and again, routine methods of oversight will not be adequate, nor will the advice of individual physicians or professional medical societies or government regulators. What is required is an intimate understanding of the nature of the research and the reliability of the results. Only with this information at hand will consumers be able to calculate potential risks and benefits to know whether to join the line outside the doctor's office, or demur.

II

The prospect of enhanced memory would be welcomed by almost everyone and would not require extraordinary individual or societal readjustments. (An expanded memory, for example, would not necessarily weaken psychological mechanisms like repression.) In attempting to fathom the "underlying biology of memory formation," various teams are trying to identify the relevant genetic components in both animals and humans.[1] One team, which is experimenting with different types of flies, has discovered that "long-term memory formation can be greatly augmented if selected single genes are activated prior to training." Moreover, it notes, "recent studies in the Drosophila, Aplysia, and Mus have been particularly suggestive of a common mechanism for long-term memory among arthropods, molluscs, and mammals."[2] Since learning itself makes a specific and physical imprint on brain circuitry, genes are likely important to the process and specific genes may release substances that aid neurons in establishing and maintaining circuitry in the regions of the brain that store information as memory. As soon as the genes and their substances are identified, methods for maintaining or improving synaptic plasticity and strengthening the activity between neurons could emerge. Hence, the team without any hesitation concludes that its finding "encourages the speculation that gene therapy, applied in humans to regions of the brain that are crucial for memory function may provide . . . interventions for repair or enhancement of memory function."[3]

No less important, if too easily overlooked, is the fact that research into memory in flies highlights the complex trade-offs that may characterize the risks and benefits of an intervention. One team tested for memory in drosophila by measuring the length of an association of a specific odor with an electric shock. The flies with better memory would sniff the odor and then avoid going into the chamber where they had earlier received the shock; those with weaker memory went right back into the chamber. The design enabled researchers to distinguish between short-term (one-day) and long-term (seven-day) memory and to begin to identify their underlying genetic components. It turns out that interventions that help extend long-term memory undercut short-term memory. More, the genes involved for memory affect other, non-memory–related, functions, including reproduction. So even at the level of the fly, there are gains and losses, with improved memory coming at the cost of reduced fertility. How this might work out in humans is unknown, but imagine the dilemma that would face an ambitious thirty-year-old stock analyst or academic.[4]

Investigators have also learned to use embryonic stem cell gene-targeting technology to produce a strain of mice with a disruption, or knockout, of a particular gene. The animal without the gene is given a series of tests to understand the functions of the now missing gene.[5] Although the technology is novel, the method is old; pioneering endocrinologists, as we have seen, castrated animals to investigate the functions of the testes. Between 1992 and 1997, some fifty articles explored the phenomenon of memory by using knockout gene technology, but almost all of them came up against a methodological problem. Since each gene has more than one function, investigators could not be certain whether a finding of reduced memory was linked to the missing gene or whether it reflected a general disability in the organism that manifested as weakened memory.[6] To be sure, research tools are becoming more refined, now able, for example, to knock out a gene later in the embryonic development and to target specific cells, not all cells, thereby reducing the possibility of general disabilities. But the field has a distance to go, including complaints from investigators that a major "shortcoming of current gene knock-out technology is not-so-infrequent cases of premature death of the mutant mice."[7]

Moving from rodents to people and from the laboratory to the clinic, memory experiments often involve patients with Alzheimer's disease (AD). For obvious reasons, memory research with humans has focused its attention here, and a future ability to enhance human memory might well result from these efforts. AD represents the progressive loss of memory and cognitive function, particularly in an older population. Although precise estimates of the total number of cases are hard to reach, it may affect some 4 million Americans. Dementia itself, which is most commonly caused by Alzheimer's, affects some 10 percent of all people aged sixty-five and almost 50 percent of all people aged eighty-five. (The percentage climbs because older cases survive and newer cases are added.) In AD, the brain undergoes several pathological changes, including the formation of amyloid, that is, dense plaques that clog intercellular space, the appearance of flame-shaped neurofibrillary tangles, and a sharp decline of neuronal activity. There is evidence, too, of cellular inflammation and a marked drop in the amount of acetylcholine in the brain, the substance that facilitates the transmission of signals across synapses.

Given the devastating consequences of AD both to the affected individual and the family, many pharmaceutical agents have been used to ameliorate the disease, but with marginal success. Prescribing cholinergic substances to offset the deficit and restore neurotransmission has not been successful. Preparations that have some efficacy in animal models have failed in humans or proved too toxic at the necessary dose levels. Velnacrine, which apparently improved memory functioning in rats, had only "modest" impact on Alzheimer's patients.[8] The FDA has approved a drug for Alzheimer's, tacrine, but since it has significant side effects (liver toxicity) and only very limited benefits, not many physicians prescribe it.[9] The list of other failures is lengthy. Drugs with antioxidant properties have no impact. Nor does extract of ginkgo, derived from the leaf of a subtropical tree and sold in health food stores as an antioxidant.[10] The record with vitamin E is no better.

Failures to identify an effective agent have intensified the search for a genetic component in AD. Although it is by no means an exclusively inherited disease, genetic influences do appear to play a role in its etiology. People who have first-degree relatives with AD are 40 times more likely than others to contract it themselves; children with affected parents have a 50 percent likelihood of getting it.[11] Although several specific genes have been implicated, the leading candidate is the (epsilon) 4 variant of apolipoprotein E (apo-E), a protein encoded by a gene on chromosome 19.[12] People who have two copies of this allele, or variant gene (2 percent of the population), are far more likely to contract the disease; by age seventy, 50 percent of this group will have it.[13] To be sure, 50 percent with the two copies will not contract it, and other considerations, such as smoking or high cholesterol, may be affecting the outcome.[14] Nevertheless, the apo-E 4 linkage persuades many investigators that future research should follow "molecular approaches that modify the effects of mutations in critical genes."[15]

Although genetic therapies have yet to emerge, an energetic research effort is under way to learn whether the apo-E 4 variant influences general cognitive performance in the normal population. Researchers have explored whether age-associated cognitive decline, which is usually considered a normal part of growing old, is actually a less severe form of Alzheimer's. If so, elderly people with the apo-E variant might be more predisposed to cognitive deficits, and were that the case, a genetic intervention might protect not only against AD but also what has heretofore been considered "normal" declines. Which brings us right to anti-aging and enhancement.

The first research results seemed to confirm the link between the gene variant and lower cognitive functioning. A study of elderly nondemented women found that subjects with the apo-E allele performed less well on memory and learning tests than women without it. Another study analyzed the genetic composition of Finnish centenarians and found that, as a group, they had few apo-E variants. However, initial findings have not held up. Subsequent research demonstrated that in non-AD persons, the presence of the apo-E variant did not bring reduced attention span, impaired language skills, or any other cognitive deficit. The allele may mark a high percentage of AD cases, but has no relevance for other conditions.[16]

In the end, the dynamic of the research, although not the immediate results, suggests just how nimbly investigations of a disease move over to enhancement. Eventually, a genetic factor may be identified that both prevents deterioration of memory and improves memory, producing an intervention capable of curing Alzheimer's and improving cognitive functioning in normal people. Then we will face the predicament of how much risk to assume for how much added memory.

III

Over the past decade, the field of behavioral genetics has gained attention for its ambitious efforts to understand and modify the genetic contribution to human behavior. Although even the most enthusiastic investigators agree that environmental contributions are important, they are confident that genetics plays a crucial role. They concede that linking complex personal interactions to genes is not easy; behavior is so rooted in family and societal contexts that forging the connections may become heavy-handed and reductionist, as though we were truly nothing other than our genes. Still, there is something commonsensical about exploring genetic influences. Aphorisms such as "like father, like son" or "the apple does not fall far from the tree" embody the idea. So does the common experience of having one's children evince behavior patterns that mirror those of grandparents they never met. Behavioral geneticists pick up on this attitude, eager to tell the story of how identical twins separated soon after birth were reunited many years later and discovered to their amazement that they both held the same kind of job, were divorced, had remarried women with the same first name, and smoked the same brand of cigarettes.

To go beyond anecdote, researchers must first devise uniform and comparable classifications of behavior, which is no simple exercise. Centuries ago, people were classified by theories of humors into one of several temperaments: they could be fiery and aggressive (the blood tipped to bile), or phlegmatic (with an excess of blood). But such groupings had fallen out of favor, at least until behavioral geneticists resurrected them. They identify five major behavioral types: people characterized by extroversion; agreeableness; conscientiousness; emotional stability; and intellectual openness.[17] Through elaborate questionnaires, they assign respondents to one or another of these categories, and then begin the search for the underlying genetic contributions.

Their strategies have been ingenious although never powerful enough to satisfy skeptics. Investigators have honed in on identical twins (monozygotic) who have been reared apart, on the supposition that similarities between them that are greater than the similarities among siblings in general indicate a genetic influence. Their next favorite group to study is identical twins reared together, whose behaviors are then compared to fraternal twins reared together. To the extent that identical twins demonstrate greater similarities than fraternal twins, the trait has a genetic component. If a trait in one identical twin appears 60 percent of the time in the other identical twin, but the same trait in one fraternal twin appears only 20 percent of the time in the sibling, then 40 percent of the trait is genetic. So if the brother of an identical homosexual twin is far more likely himself to be homosexual than the brother of a fraternal twin, then homosexuality has a genetic component.

Researchers also study adopted children to learn whether their personality traits resemble more their adoptive parents (which speaks to the environment) or their biological parents (which speaks to genetic inheritance). Perhaps most controversially, researchers are analyzing children and adults with special behavioral features, some desired (musical aptitude), some undesired (mental illness, criminal behavior), to see whether they run in families. Were children of schizophrenic parents or incarcerated parents more likely to suffer the illness or be convicted than others, genetics might be a factor in the etiology of the condition.

The ultimate task of the research is by far the most difficult: to locate the gene or, more likely, the several or many genes responsible for the behavior. Investigators must find the allele that is common to schizophrenics or to extroverts. Then they have to figure out what this genetic variation does biologically and how it might be manipulated to do more or to do less or to do otherwise. The whole enterprise may seem fantastic, but imagine if an allele could be tweaked to embolden shy people or tame aggressive people.

What has been accomplished so far? Researchers believe that they have identified a number of personality characteristics with genetic bases. One investigator finds that "extroversion," for example, correlates to .68 for monozygotic twins reared apart, .55 for such twins reared together, and .20 for biological siblings, thereby indicating that "individuals who share genes are alike in personality regardless of how they are reared."[18] To be sure, the environment, and the "non-shared environment," that is, the social settings that one sibling experiences but not the other, exert a powerful influence on an individual. But if the trait did not have a genetic base, why is it more common to monozygotic twins reared apart than to biological siblings?

Other researchers have searched for the genetic basis for "novelty seeking." Although they report correlations with a gene known as D4DR, which helps regulate dopamine (the powerful neurotransmitter whose absence causes Parkinson's disease), their own calculations suggest that the genetic influence affects only a small proportion of people with the trait.[19] "D4DR," concludes one researcher, "accounts for roughly 10 percent of the genetic variance, as might be expected if there are 10 or so genes for this complex, normally distributed trait."[20] No wonder that to date "modest" is the descriptor usually attached to even positive finding of behavioral genetics.

Although the infant science may grow to maturity, fundamental problems complicate its research—and before succumbing to its allure or interventions, it is important to understand them. First, the personality categories lack internal consistency. The "extroversion" type includes people who are gregarious, social, dominant, and adventurous. But these traits are at odds with one another—gregarious may conflict with dominance, social with adventurous—giving the category a grab-bag quality. This is even more true for "novelty seeking," which lumps together such behaviors as exploratory, fickle, impulsive, extravagant, and quick-tempered; those inclined to be exploratory might not be, or wish to be, extravagant and quick-tempered, traits that would certainly impede effective explorations.[21]

Second, twin studies that rely on the fact of being "reared apart" may be misleading. Although the twins may not be residing with the same family, they may be in very similar families—social welfare agencies typically place children with adoptive families that closely resemble their biological families in color, religion, social class, and general interests. Moreover, twins who have a stake in being twins—getting psychological payoffs for the status—may exaggerate the qualities that they share, or even the extent to which they have been reared apart.[22]

Third, little progress has been made to date in linking personality traits to specific genes or alleles. The likelihood is that when biologic bases for behavior are identified, there will be multiple genes interacting in different ways with one another and with the environment to produce the results. To carry out what is called a "brute force scan" of the entire genome, casting a very large net to find a putative gene, is difficult and expensive. The alternative, to use families known to have the trait in order to identify it, which works well in a one-to-one, gene-disease model, is less effective with behavioral traits. It is far more difficult, really almost impossible, to identify and locate an extroverted family or a novelty-seeking family than it is to identify such an individual. Yet, searching for genetic similarities among single individuals who scored high on a screening test is too broad-gauged an approach to genetic screening. To make research matters worse, when the trait under investigation is a negative or stigmatized one, as may well be the case for novelty-seeking and surely is for criminal behavior, research faces extraordinary barriers. Informants may lie about their behavior, or communities, fearful of stigma, may protest vigorously against the project. In sum, the promise of behavioral genetics to enhance performance is still very much in the future, and when potential interventions do emerge, they will be particularly chancy and problematic.

IV

These many complications notwithstanding, the quest to understand the genetic bases for behavior is certain to continue, driven by the desire to cure illness, especially mental illness. The best example is schizophrenia.[23] This devastating disease disproportionately strikes young people as they enter adulthood. Given the special symptoms that mark it, including hallucinations, delusions, speech defects, and flattened affect, many investigators are persuaded that it is physiological, not psychological, in origin, reflecting deficiencies in "neural connectivity." Schizophrenia becomes a disease of the brain that manifests itself as a disease of the mind, and an excellent candidate for genetic study.

Researchers have long recognized a pattern of family inheritance in the occurrence of the disease. Only 1 percent of the population but 13 percent of the children of a schizophrenic parent are at risk; when both parents are schizophrenic, the risk climbs to 46 percent. So, too, 48 percent of identical twins with a sibling who is schizophrenic will themselves contract the disease.[24] Epidemiological research using the exceptionally complete Danish medical and population records was able to identify 2,669 cases of schizophrenia in a country of some 1.75 million people; the odds that any given person would contract schizophrenia were far greater if mother, father, or sibling had it than for others. To be

sure, most schizophrenics did not have an affected parent and even
among identical twins fully half did not contract the illness after their
sibling did. The Danish data also pointed to the effects of the environment. People born in cities were more likely than their rural counterparts to suffer the disease. Children whose mothers had a viral infection
during pregnancy were more likely to become schizophrenic.[25] In all,
schizophrenia is by no means an exclusively genetic disease, but it probably has significant genetic components.

Despite considerable effort, researchers have not been able to identify the gene(s) that contribute to the etiology of the illness. They have
searched for alleles on more than half of 23 chromosomes, but still do
not know which play a crucial role. What confounds the search? For
one, defining the characteristics of schizophrenia (the phenotype to be
studied) turns out to be very complicated. The psychiatric diagnostic
manuals (*DSM* III and *DSM* IV) are useful starting points but are not
sufficient to the task. The symptoms of the disease are so wide-ranging
that *DSM* III and IV have to set out four different categories, running
the gamut from a narrow to a very broad definition. Manic-depressive
behavior fits the narrow category; alcoholism and anxiety join in the
broadest category.[26] Moreover, the manuals recognize that some schizophrenics will manifest the disease with reduced emotional affect and
apathy, while others will be delusional and hallucinating.

The breadth of symptoms does not trouble a practicing psychiatrist
who writes a prescription for a particular patient. Most of the drugs
are broad-acting, covering a wide range of symptoms. But genetic
researchers are in a more difficult position. It is entirely possible, even
likely, that each behavioral pattern has its own special or only partly
overlapping genetic roots. Schizophrenia may not be, genetically speaking, one, but several diseases. So, if one research team works on a study
population that is mostly in one category (say, manic) and another team
uses a study population from another category (say, alcoholic or delusional), the likelihood that the teams will confirm each other's genetic
findings is reduced. Moreover, the difficulties in identifying the right
genes are exacerbated since schizophrenia is almost certainly not a single gene defect, like Huntington's disease; small variations among many
genes may add up to a predisposition to the disease, which may or may

not then occur, depending on the particular environment. To circumvent some of these problems, investigators are attempting to identify physical traits that accompany all variants of schizophrenia, such as abnormalities in gait and unusual eye movements. But the effort is highly preliminary and may further complicate investigators' tasks. Now they will have to identify what may turn out to be separate although associated genetic determinants of both a physical and a mental disease.

In the meanwhile, genetic research findings regularly contradict one another. One group announces that it has located a promising lead on a section of chromosome 13, which some investigators then confirm and others disconfirm. Another team identifies a relevant region on chromosomes 5 and 10, but others disagree. There have been several positive findings for regions on chromosomes 6, 8, and 22, but since a region, or "locus," on a chromosome involves hundreds of genes, it is not easy to narrow the finding even "on these potentially promising regions."[27] For the moment, "no specific gene in any of the regions has been identified, let alone one identified with a specific neurochemical defect."[28] No interventions are in sight, either for cures or for enhancements. But optimists believe that breakthroughs on both fronts are sure to occur, so the question of which team and which findings to trust will sooner or later be upon us.

V

Research into the genetics of alcoholism presents even more intriguing enhancement possibilities. Many twin and adoption studies point to a hereditary component in alcoholism. If one identical twin is an alcoholic, then the likelihood that the other will be is greatly increased in comparison both to fraternal twins (when one is an alcoholic) and to the general population. The finding holds as well when the identical twins are reared apart.[29] So too, children (particularly male) of an alcoholic father are 2 to 10 times more likely than others to become alcoholics. But almost all of the problems endemic to schizophrenia research reappear here as well. To identify the relevant genes, researchers must delineate specific behavioral characteristics within the general trait of alcoholism, identify subsets of the population that manifest them, review the genetic characteristics of these subsets to find the alleles that might be common to them and associated with alcoholism, and finally, understand the biological mechanisms at work and learn to modify them.

Each step poses a myriad of problems. Alcoholism is a far more elusive category even than schizophrenia, representing, as one investigator notes, "one of the most complex [syndromes] known."[30] Described as a psychiatric illness in *DSM* IV, alcoholism manifests itself in so many different ways that researchers are hard-pressed to delineate discrete populations for study. It varies by age of onset (from adolescence to adulthood), by amount of drinking (hard drinkers to very hard drinkers), and by the severity of hangovers (moderate to heavy). There are differences by sex and by group (Japanese versus Americans, African-Americans versus Caucasians). Sometimes, but not always, alcoholism correlates with smoking, illicit drug use, and violence. So a vicious drunk may not be the same as a happy drunk. A drug-abusing drunk may not be the same as one who only drinks. To the extent that such differences matter, the relevant gene(s) become more elusive.

Researchers looking for uniform patterns above this behavioral diversity have uncovered unique brain wave patterns among alcoholics, along with particular visual and spatial defects and low blood platelet activity.[31] They also find metabolic deficiencies (the liver is unable to process the alcohol) and neurological deficiencies (lowered serotonin or endorphin levels). But the question remains whether these changes are the causes or the effects of alcoholism. Does a lowered serotonin level follow on hard drinking or encourage hard drinking? To answer the questions would require identifying potential alcoholics in advance, measuring their physiological functioning, and then returning to study them after they have, or have not, become hard drinkers. It is not an impossible design, but it is certainly not a simple or efficient one.

All these considerations affect how researchers conduct their genetic studies. When selecting a cohort of alcoholics to investigate, they must decide whether to rely on a personality questionnaire, a psychiatric interview, behavioral criteria, or a neurological, hematological, or gastroenterological examination. They have to choose between early-onset alcoholics and late-onset alcoholics and alcoholics with or without histories of other types of substance abuse. Given the range of choices, different teams invariably make different selections, which helps explain why genes identified by one study do not turn up in another. The situation is tailor-made for divergent findings.

One genetic marker for alcoholism that has attracted some support among researchers is the allele ALDH2-2 on chromosome 4. Those who carry this variant metabolize alcohol much more slowly than normal and, therefore, manifest such alcohol-related symptoms as a flushed face and a racing heart. Because it generates these unpleasant side effects, the allele is probably protective against alcoholism and is found disproportionately among Asian and Jewish populations, both groups with low levels of alcoholism.[32]

Findings about another allele, DRD2 A1, located near the dopamine receptors in the brain, have spurred much more controversy. Ernest Noble, a psychiatrist at UCLA, and Kenneth Blum, a pharmacologist at the University of Texas Health Center at San Antonio, reported in *JAMA* in 1990, that after examining DNA from thirty-five deceased alcoholics, they found the DRD2 allele in 69 percent of the sample, compared to 20 percent in a normal control group. What made this finding so interesting is that people with this very same allele have abnormally low levels of dopamine. Accordingly, Noble and Blum contended that since alcohol is known to increase dopamine levels in the brain, alcoholism may well be a gene-based response to a shortage of dopamine. The alcoholic is drinking because his body is trying to raise its dopamine levels. Extending this argument further, they suggest that DRD2 A1 is actually not itself an allele that causes alcoholism but rather is a "reward gene" that plays a crucial role in many types of addictive behavior. Because the allele reduces dopamine levels, it triggers a variety of responses, all seeking to raise the levels, and the responses can take the form of food abuse and drug abuse, as well as alcohol abuse. Thus, we may have at hand "a new paradigm in our understanding of the genetic basis of addictive-compulsive behaviors."[33]

As would be expected, other investigators immediately attempted to replicate these findings but, with a single exception, could not. One team considered the argument a "castle in the air," another was highly doubtful that the allele was actually a "trait marker for susceptibility to alcoholism."[34] But Noble did not back off. He insisted that other investigators had not used the appropriate cohort of alcoholics and control groups. For example, they had wrongly excluded severe alcoholics and had not tested the controls for other addictions that might account for the presence of the same allele. Moreover, he cited findings that treating alcoholics with bromocriptine, a substance that raises the body's level of dopamine, was proving an effective antidote to alcoholism; among those with the allele, the group receiving bromocriptine showed "the greatest and most significant decreases in craving and anxiety."[35] Noble was also encouraged by the findings of an NIH research team that, using a new scanning device, reported that obese people had fewer dopamine receptors than others and concluded that overeaters were trying to increase their dopamine levels and produce "feelings of satisfaction and pleasure." In light of all this, Noble suggested that "subjects with this [DRD2] allele may compensate for the deficiency of their dopaminergic system by the use of alcohol and other substances, agents known to increase brain dopamine levels. Stimulation by dopamine of A1 allelic subjects' fewer D2 dopamine receptors could provide enhanced feelings of reward and pleasure."[36]

If Noble is right, the prospects for genetic enhancement have escalated. Suppose investigators identify a gene or genes that could be manipulated to stimulate dopamine receptors and dopamine levels and thereby increase feelings of pleasure and reward. Such an intervention would likely win FDA approval as a means to treat alcoholism, obesity, and substance use. In short order, however, it would be prescribed off-label to elevate feelings of pleasure in a normal cohort. Investigators' curiosity, physicians' readiness to make their patients happy or happier, company marketing strategies, and people's readiness to experiment with new agents will ensure that even before the data are clear or potential side effects recognized, the technology will be used.

VI

Of all the enhancement possibilities, none has greater potential for transforming individual lives and social relationships than a dramatic extension of life expectancy. Some geneticists confidently predict that new knowledge and techniques will be able to double the life span, not by conquering the leading causes of death, but by understanding the cellular basis of aging.

The prospect of realizing this agenda has dismayed not only a variety of social commentators but also prominent cell researchers. They fear that the changes would be unnatural, burdening the environment, sparking an intergenerational warfare for power and resources, and stifling innovation. Overpopulation might well lead to state-coerced birth control. The hundred-plus-year group would have such an iron grip on resources that innovation in political, economic, and cultural life would be drastically reduced. Even from the individual's perspective, longevity might be a curse. One critic quotes a 1922 play, *The Makropoulos Secret*, written by Karel Capek, (famous for his *R.U.R.* robot story), in which the heroine, Emilia, having swallowed the now lost recipe of longevity concocted by her father in the sixteenth century, declares in her 337th year that her problem is "Boredom. No, it isn't even boredom. . . . You people have no name for it. . . . One cannot stand it. For 100 years, one can go on. But then . . . one's soul dies. . . . No one can love for 300 years—it cannot last."[37] Another opponent, Leonard Hayflick, whose original findings on cellular division lay the foundation for the cutting-edge work in the genetics of aging, wants nothing to do with promoting longevity. "I do not believe that our goal should be to tamper with the fundamental processes of aging, even if that were possible," he recently remarked. "The problems that would be created far outweigh any potential benefits. Indeed, I cannot imagine any scenario where tampering with the aging process would be beneficial. . . . To arrest or stop the aging process will not accrue to our benefit or that of society."[38]

Other investigators, however, are impatient with these objections. After all, to a thirty-year-old in 1800, the prospect of people living into their eighties and nineties would have seemed no less perverse. And by what standard is it wrong to extend the life of those already on the planet even if it meant creating fewer children? The 140-year-olds would bring their added wisdom and experience to the service of mankind. The likelihood is also strong that the technologies and knowledge that doubled longevity would ensure that the elderly would be healthy enough to remain employed, not idle in a nursing home. Personal and social readjustments will be necessary, but it is not beyond our ability to learn how to function as great-great-grandparents.

Longevity, as in the other fields we have explored, probably has a genetic component. Epidemiological data suggest that just as children of alcoholics are more likely to become alcoholics, siblings of centenarians are far more likely to live longer than the general population; indeed they are 17 times more likely than others to become centenarians themselves. Animal studies also indicate a genetic contribution. In one simple and convincing experiment, researchers selected the longest-lived mice from a cohort and then mated them with other long-lived mice and eventually created a breed that had twice the longevity of ordinary mice.[39] In another experiment, investigators working with both the nematode worm (selected for a particular genetic mutation) and the fruit fly (selected for later fertility) were able to breed strains that had twice the normal species life span. In the case of the fruit fly, longevity was accompanied by other "functional enhancements," including resistance to stress and "increased locomotor capacity at later ages."[40] Not deterred by the biological distance from flies and worms to humans, these investigators confidently concluded that "we can now legitimately address the question of the postponement of human aging in a practical manner."[41]

In fact, mice have provided the most dramatic example of an intervention that actually increases longevity. Adult mice placed on caloric restriction (CR), following diets 25 percent to 50 percent lower than normal, outlive and outperform control groups. Why CR exerts this effect is not understood; it may improve DNA repair capacity or alter gene expression or reduce oxidative stress or lower metabolism. And even though hundreds of genes may be involved in the changes, the findings have been called "among the most important ever obtained in gerontology. . . . They constitute some of the best evidence available concerning the postponement of aging in mammals."[42]

Other longevity research takes as its starting point genetic diseases that cause premature aging and death. Werner syndrome affects the one person in 25 million people who has two copies of a defective gene, WRN. Because of this mutation, they show signs of aging in their early twenties, including balding and loss of skin elasticity; in their thirties, they suffer diabetes and osteoporosis. They die on average in their late forties, usually from cancer or atherosclerosis. Another premature aging disease is Hutchinson-Gilford progeria, in which very young children suffer skeletal damage and cardiac disease, giving them an average life span of fifteen years. By studying these two diseases, geneticists hope to identify the particular gene(s) at fault for premature aging and for atherosclerosis; when they understand the mechanisms, they will be better able to grasp the role of the many other genes that affect aging. One team has already identified a genetic variation that might explain why patients with progeria have a greater incidence of myocardial infarctions, and the finding might have relevance to heart disease in the general population.

Although researching the genetics of early-aging diseases seems a sensible strategy, not everyone agrees with it. Rare diseases may involve completely idiosyncratic genetic defects and not serve as useful models. In the case of myocardial infarctions among progeria patients, for example, other teams have not replicated the identification of a particular genetic variation. Nevertheless, the study of these rare diseases provides a telling example of how research with a therapeutic goal can easily have profound enhancement consequences. As investigators learn to prevent premature aging through a genetic fix, they may also learn to prevent "normal" aging through a genetic fix—at which point we may enter a period of dramatically increased longevity.

But only "may enter" because genes may not be all-powerful. For one, environmental influences cannot be ignored. Longevity rates differ by socioeconomic class and by race; the wealthy live longer than the poor, whites longer than blacks. For another, genetic advantages in promoting longevity might not pass from generation to generation; life span *after* reaching the age of reproduction falls outside the dynamic of natural selection. As Darwin explained, the process favors only those genetic advantages that enhance reproductive capacity, not duration of survival afterward. There might even be a conflict between reproductive capacity and longevity. According to the concept of antagonistic pleiotropy, the attributes that contribute to effective early survival and reproductive capacity up to age thirty, including heightened immune responses and high levels of such hormones as testosterone and estrogen, may, twenty or thirty years later, turn out to be disadvantageous to good health; a heightened anti-immune response might cause autoimmune disease, and a high level of estrogen, cancer. By the same token, natural selection might not encourage the development of highly efficient and continuous cellular repair mechanisms were these mechanisms to inhibit cellular growth in the first thirty years of life and, thereby, reduce reproductive potential.

Another daunting and, as we have seen, familiar challenge facing researchers is to move from hypothesizing genetic contributions to longevity to understanding the processes by which genes affect longevity. The most promising approaches follow on the findings of Leonard Hayflick. Hayflick discovered that the number of times a cell will divide has an upper limit; after fifty divisions, the cell falls into disrepair and soon dies. This limit is not a response to a biological clock—elapsed time is not the determining feature—but to a biological counter. After fifty divisions, in what has become known as the "Hayflick limit," cells became mortal. There are a few exceptions. Cancer cells keep dividing and dividing, and in that sense are immortal, and so are stem cells. But all others are not.[43]

What sets the counter? How do cells know when they have reached fifty divisions? The answer was found in the end of the chromosome, the telomere, as it is called, whose purpose is to keep the twisted ends of the double helix of the chromosome from attaching to each other and blocking DNA division and replication. Each division of the cell causes a progressive shortening of the telomere, and the shortening eventually disrupts normal DNA replication. After fifty divisions the telomere becomes so short that the cell can no longer divide effectively; some of its genes are then turned off, others turned on, including some oncogenes (which may turn a normal cell into a tumor cell), and still others disarranged so that they send out disruptive messages. In Werner syndrome, for example, telomeres allow cells to divide at only half the rate of normal cells, which may help explain premature aging.

Once the function of telomeres became clear, the next question was whether some methods might be discovered to take cellular divisions beyond fifty. Researchers plunged in. The question was never whether to do it—concerns about the social impact of longevity did not cause a ripple—but how to do it. How could cellular division be increased and cellular aging prevented without turning the newly invigorated cells into cancer cells? One answer came in 1998 from a team made up of investigators from the Geron Corporation and Texas Southwestern Medical Center. They inserted a newly cloned enzyme, hTRT, or telomerase, into human cells and these cells then went on to divide past the Hayflick limit by no fewer than thirty-six added doublings. Apparently, decreasing amounts of telomerase enzyme restricted cellular division; if the amount of the enzyme was increased, the cell would continue to divide. The team reported finding no gross changes in the proliferating cells and no signs of cancerous cells. They were not surprised—stem cells proliferated indefinitely without becoming cancer cells.

The team addressed the significance of the finding by moving immediately from cell to organism, and, curiously, from the active voice to the passive voice, as though the conclusion was too startling to consider. "Cellular senescence is believed to contribute to multiple conditions in the elderly that could in principle be remedied by life-span expansion." These remedies could correct for loss of skin elasticity, age-related deterioration of optical cells, and atherosclerosis. The report ended with an obvious understatement that echoed the famous line that closed the Watson-Crick paper on DNA: The findings had "important implications for biological research, the pharmaceutical industry, and medicine."[44]

As would be expected, the report attracted substantial attention. Writing in *JAMA,* Michael Fossel, editor of the *Journal of Anti-Aging* (and, as his disclosure form noted, a stockholder in Geron), first explained the "startling" research that took cells beyond the Hayflick limit.[45] Then, he, too, jumped from cell to organism, extending the diseases that the new method might cure to arthritis, immune deficiencies, and Alzheimer's disease, and minimizing the risks that might accompany such an intervention. There was "no evidence yet that telomerase expression per se causes malignant transformation."[46] (The "yet" was Fossel's hedge against future findings.) And he, too, closed his review awkwardly but portentously: Medicine was advancing in its "potential to treat human disease, to alleviate patient suffering, and—raising the possibility 'in proportion to its dignity'—that we may alter 'the thread of life itself.' "[47]

Although other teams soon confirmed the enzyme's ability to extend cellular division, fundamental disagreements erupted over the question of risk, specifically whether the addition of telomerase reduced the cell's ability to protect itself from becoming a runaway cancer cell. In January 1999, the Texas team confirmed its earlier finding that using telomerase to double cell life span "does not lead to alternated patterns of growth typical of cancer cells."[48] The process "does not bypass cell-cycle induced checkpoint controls and does not lead to genomic instability."[49] But then, in June 2000, another team found otherwise. Although it readily acknowledged the promise of "cell-based therapies by allowing indefinite expansion of normal human cells without damaging their genomes," its own studies of cells that surpassed the Hayflick limit uncovered an increased presence of a particular oncogene, one whose activation "occurs in a wide variety of tumour types."[50] In fact, adding telomerase to the cell "has little, if any, immortalizing potential until the . . . tumour suppressing gene has been inactivated." In other words, the very effectiveness of the enzyme depended on its undermining the cell's protection against cancer growth. The team's conclusions, therefore, were far more guarded: "Expansion of normal human cells for therapeutic purposes must be approached with caution."[51]

Telomerase research actually reversed the traditional line of influence that takes a cure on to enhancement—in this case, enhancement research fed back into cure. Even as investigators were exploring the role of telomerase in extending longevity, others were examining telomerase as a potential cancer therapy. Since almost all cancer cells reactivated telomerase, the enzyme might be fueling malignant growth. If it were possible to inhibit the production of telomerase, the growth of tumor cells might be arrested. It was a promising but very risky strategy: restrict cell growth and so increase the prospect of cell death in order to curtail cancer cell immortality. Whatever the eventual outcome of these approaches, telomerase, like so much else in the enhancement arsenal, is a double-edged sword.

As novel as these cellular findings are, their implications follow the analytic path that we have marked out before. First, no matter how vocal critics may be about the deleterious consequences of increased longevity, the scientific work is far too fascinating and consequential for investigators to abandon it. Second, even were there a systematic effort to rein in laboratories, it would be impossible to distinguish anti-aging research from anti-disease research. Since telomerase and telomeres may hold a key to the cure of cancer, research is certain to move ahead.

Third, the new telomere technology is also likely to present an exquisite risk-benefit calculus. Since aging occurs gradually over time, it may be too late to intervene at age sixty or seventy. By then, the cellular damage may have occurred. The right time to act may be at age thirty or forty, when the subject is healthy, in the prime of life, with the immune system at its peak, the DNA repair system working well, and the body still free of accumulated cellular defects. But consider the risks of experimenting with very young and healthy subjects. An intervention that might extend longevity might also cause disability or premature death. How should one balance the possibility of immediate risks with benefits that might (or might not) appear decades later? It might prove feasible to test the new technologies on the very elderly, willing to volunteer because they think they have less to lose. It might even be possible to devise surrogate markers for mortality so that the effectiveness of the intervention could be known more quickly. But there is no certainty that such solutions will be feasible.

These problems might give even daring investigators, ambitious biotech companies, and adventurous consumers pause. But if the past is a guide, there is certain to be someone ready to do the research, someone to fund it, and someone ready to serve as the first subject. Yes, there will be risks—but just imagine enjoying the benefits of an extra seventy years.

VII

We close where we began, with a recognition of the ongoing, in fact, intensifying, interest in the promise and perils of medical enhancement. Even without any stunning breakthroughs in knowledge or technology, the first six months of 2003 witnessed a remarkable outpouring of books, both fiction and non-fiction; articles; and newspaper stories on the subject. The stance of some authors is predictable. Scientists, for example, have difficulty understanding what all the fuss is about. James Watson took the occasion of the Fiftieth Anniversary of his and Francis

Crick's publication of the double helix structure of DNA to argue in favor of the pursuit of enhancement. Elevating genes over environment, he insists that rather than invest in extra tutorials for slow learners, we should be exploring how to manipulate thair genetic inheritance to bring them up to speed. Like so many of his fellow investigators past and present, he is thoroughly impatient with the idea that nature knows best. Watson speculates that were it not for the Nazi experience, we would have few doubts about the wisdom of genetic enhancement. Indeed, that very experience prompts him to observe: "Do we dare restrain our own research community, especially in light of the fact that other research communities, less trustworthy than ours, might be making progress?" His answer is absolutely not.[52]

Fictional accounts, like those by Margaret Atwood, take a very different posture, fearful that enhancement technologies will profoundly upset the balance of nature. Atwood's latest novel, *Oryx and Crake*, follows in the spirit of her earlier *The Handmaid's Tale*, and in the still earlier traditon of Mary Shelley. Her protagonist's father helped to "engineer the Methuselah Mouse as part of Operation Immortality," worked on the "pigoon project," to grow multiple organs for human transplantation, and to "create a genuine start-over skin that would be wrinkle- and blemish-free."[53] (His wife objected to the projects: "You're interfering with the building blocks of life. It's immoral. It's . . . sacrilegeous.")[54] But worse was yet to come. The son, Jimmy/Snowman, who is among the last of human kind, lives in a wasteland. The devastation was the result of a viral epidemic that was spread by the anti-hero Crake, who promoted pills that promised users protection against all sexually transmitted diseases, an unlimited supply of libido and sexual prowess, and prolonged youth. It delivered, on a worldwide scale, death. Along the way Atwood lampoons institutes devoted to "Anoo-Yoo" and "Rejoov," as well as "pills that make you fatter, thinner, hairier, balder, whiter, browner, blacker, yellower, sexier, and happier."[55] We pursue enhancement at our collective peril.

Bioethicists such as Carl Elliott, in *Better Than Well*, share a visceral unease about enhancement. It "rubs me the wrong way," Elliott declares; it is "too close to fakery."[56] He finds a paradoxical quality in consumers' responses to the technology. Although they describe it in terms of fulfilling a personal sense of identity, they are actually allowing the opinion of others to determine their sense of self. A "social mirror," not their own judgments, holds sway. Although you may explain your preference for plastic surgery in terms of satisfying your own desires, "you feel different because of the way other people perceive you."[57] Ultimately Elliott cannot make his peace with "the fragility of selves that depend so intimately on the good opinion of others for their survival."[58] So if Atwood is worried about the threat of enhancement to the well-being of all of us, Elliott is concerned with its threat to the authenticity of each of us.

All the while, the media cover enhancement closely. They highlight purported innovations in laboratories and clinical practice. The news section along with the style and personal sections devote considerable time and space to the latest developments in mouse memory and longevity, as well as to cosmetic surgery and liposuction techniques. At the same time, the media also issue cautions based on new research findings. It is difficult to know what readers take away from these mixed messages, but there is, unavoidably, a spasmodic quality to the coverage. Yes today, no tomorrow—with little consistency of perspective within, let alone among, the presentations.

Perhaps most disappointing, medicine itself has provided very little leadership. It is not surprising that drug companies will be less interested in educating consumers to be skeptical about new products than in persuading them to have their physicians prescribe the latest pill. But surely medicine has an obligation to be more proactive and diligent. There are self-correcting mechanisms in research; eventually, investigators learn whether a particular intervention does more harm than good. But *eventually*, as with estrogen, can mean decades of conflicting findings about risks and benefits. To be sure, some specialty societies issue practice guidelines. Even then, however, as the use of testosterone and growth hormone as anti-aging compounds reveals, medical organizations make almost no effort to discipline physicians who violate these guidelines, some prescribing these compounds through the Internet without so much as examining the patient. Nor will they curb physicians who present themselves as anti-aging specialists when practically every expert agrees that there are no effective anti-aging medicines. Should a state legislature attempt to regulate medical practice, for example, mandating greater oversight in office-based surgery, physicians together with medical societies will strenuously protest. But rarely does the medical community step in to fill the vacuum.

From our perspective, the real dangers of enhancement do not lie in our souls (as with Elliott) or in our ambitions (as per Atwood) or in our history (as per Watson) but in our enthusiasms and credulity. We do not believe that enhancement will necessarily violate nature, subvert our humanity and dignity, or undermine social order. What the technologies do represent is a test of the outer limits of allowing science to set its own agenda, of allowing happiness to drive clinical care, of allowing profit motives almost unbounded license, and allowing individuals to exercise autonomy and choice. Setting boundaries is an exceptionally difficult task. Let moralistic minorities abridge science? Restrict doctors from responding to emotional pain? Substitute state collectivism for the profit motive? Enshrine paternalism over choice? None of these are attractive, or even feasible, alternatives. What remains is the possibility, no more than that, of consumers making decisions wisely, taking account of both self-interest and societal interest. They will need to be at once wary and restrained to minimize personal and collective risks, and open and generous to democratize benefits. If we have encouraged and contributed to such an enterprise, we will have accomplished our task.

Notes

Introduction

1. H. J. Muller, *Out of the Night: A Biologist's View of the Future* (New York: Vanguard Press, 1935), 125.

2. Joshua Lederberg, "Viruses and Humankind," in *Emerging Viruses,* ed. Stephen S. Morse (New York: Oxford University Press, 1993), 3–9.

3. National Institutes of Health Gene Therapy Policy Conference, "Human Gene Transfer: Beyond Life-Threatening Disease," Bethesda, Maryland, September 11, 1997. See also Eric T. Juengst, "Concepts of Disease After the Human Genome Project," in *Ethical Issues in Health Care on the Frontiers of the Twenty-First Century,* ed. Stephen Wear, James J. Bono et al., 125–52 (Dordrecht; Boston: Kluwer Academic Publishers, 2000); David B. Resnick, "The Moral Significance of the Therapy-Enhancement Distinction in Human Genetics," *Cambridge Quarterly of Healthcare Ethics* 9 (2000): 365–77.

4. Leon Kass, "The Wisdom of Repugnance," *New Republic* 298 (June 2, 1997): 18.

5. Francis Fukuyama, *Our Posthuman Future* (New York: Farrar, Straus and Giroux, 2002). On the other hand, William Haseltine, head of a new biotechnology company, comments: "We are redefining what it is to be a human—taking a new measure of man. We add to the classical and neo-classical view of the human body as a collection of organs, tissues and cells, the notion that [the] human body can also be described as a collection of genes and the protein each gene specifies," in Howard Gardner, Mihaly Csikszentmihalyi et al., *Good Work: When Excellence and Ethics Meet* (New York: Basic Books, 2001), 81–82.

6. William Kristol and Eric Cohen, "Should Human Cloning Be Allowed?," *Wall Street Journal,* December 5, 2001.

7. Lee Silver, *Remaking Eden* (New York: Avon Books, 1997), 4–7. See also Maxwell Melhman, "How Will We Regulate Genetic Enhancement," *Wake Forest Law Review* 67 (1999): 671–714.

8. In the extensive literature on enhancement, we found most useful: LeRoy Walters and Julie Gage Palmer, *The Ethics of Human Gene Therapy* (New York:

Oxford University Press, 1997); Philip Kitcher, *The Lives to Come: The Genetic Revolution and Human Possibilities* (New York: Simon & Schuster, 1996). See also: George J. Annas, "The Man on the Moon, Immortality, and Other Millennial Myths: The Prospects and Perils of Human Genetic Engineering," *Emory Law Journal* 49 (2000): 753–82; William Gardner, "Can Human Genetic Research Be Prohibited?," *The Journal of Medicine and Philosophy* 20 (1995): 65–84; James L. McGaugh, "Enhancing Cognitive Performance," *Southern California Law Review* 65 (1991): 383–95; Thomas H. Murray, "Assessing Genetic Technologies: Two Ethical Issues," *International Journal of Technology Assessment in Health Care* 10 (1994): 573–82; Michael H. Shapiro, "The Technology of Perfection: Performance Enhancement and the Control of Attributes," *Southern California Law Review* 65 (1991): 11–113.

ONE   Penetrating Nature

1. Claude Bernard, *An Introduction to the Study of Experimental Medicine*, trans. Henry Copley Greene (New York: Macmillan, 1927).

2. Bernard, *Experimental Medicine*, 6.

3. Bernard, *Experimental Medicine*, 9.

4. On Bernard's contribution to the ethics of human experimentation, see David J. Rothman, *Strangers at the Bedside: A History of How Law and Bioethics Transformed Medical Decision Making* (New York: Basic Books, 1991).

5. Bernard, *Experimental Medicine*, 18.

6. Quoted in William Coleman, "The Cognitive Basis of the Discipline: Claude Bernard on Physiology," *Isis* 76 (1985): 54, 56.

7. Bernard, *Experimental Medicine*, 103.

8. Janet Browne, *Charles Darwin: The Power of Place*, vol. 2 (New York: Alfred A. Knopf, 2002), 56.

9. Gillian Beer, *Darwin's Plots* (London: Routledge & Kegan Paul, 1983), 10, 21.

10. Quoted in Charles Rasmussen, *Jacques Loeb: His Science and Social Activism and Their Philosophical Foundations*, Memoirs of the American Philosophical Society, vol. 229 (Philadelphia: American Philosophical Society, 1998), 112, 115. See also Philip J. Pauly, *Controlling Life: Jacques Loeb and the Engineering Ideal in Biology* (New York: Oxford University Press, 1987).

11. J. B. S. Haldane, *Daedalus, or Science and the Future* (New York: Dutton, 1924).

12. Haldane, *Daedalus*, 5, 6.

13. Haldane, *Daedalus*, 44.

14. Haldane, *Daedalus*, 19, 54, 58, 73.

15. Haldane, *Daedalus*, 70–71.

16. Haldane, *Daedalus*, 74.

17. Haldane, *Daedalus*, 66–67.

18. Haldane, *Daedalus*, 80, 82.

19. Robert M. Philmus and David Y. Hughes, eds., *H. G. Wells: Early Writing in Science and Science Fiction* (Berkeley: University of California Press, 1975), 36. The essay was originally published January 19, 1895.

20. Philmus, *H. G. Wells*, 38.

21. Philmus, *H. G. Wells*, 39.

22. H. J. Muller, *Out of the Night: A Biologist's View of the Future* (New York: Vanguard Press, 1935), 179.

23. H. J. Muller, "The Guidance of Human Evolution," *Perspectives in Biology and Medicine* 3, no. 1 (Autumn 1959): 1, 8, 42.

24. As quoted in Daniel J. Kevles, *In the Name of Eugenics: Genetics and the Uses of Human Heredity* (New York: Alfred A. Knopf, 1985), 48.

25. Quoted in Kevles, *In the Name of Eugenics*, 127.

26. On the history of physiology, especially in England, see Gerald L. Geison, *Michael Foster and the Cambridge School of Physiology* (Princeton: Princeton University Press, 1978), especially ch. 11; for the American story, see W. Bryce Fye, *The Development of American Physiology* (Baltimore: Johns Hopkins University Press, 1987).

27. Bernard, *Experimental Medicine*, 107.

28. Bernard, *Experimental Medicine*, 98.

29. Bernard, *Experimental Medicine*, 99.

30. Bernard, *Experimental Medicine*, 101.

31. Bernard, *Experimental Medicine*, 103.

32. Quoted in Fye, *American Physiology*, 3.

33. Quoted in Max Thorek, "Studies in the Technic and Clinical Application of Sex Gland Transplantation," *Journal of the Michigan State Medical Society* 23 (1924): 532.

34. Ernest Starling, *Principles of Human Physiology* (Philadelphia: Lea Febiger, 1912) 1318–19. See also Victor Medvei, *A History of Endocrinology* (Lancaster: MTP Press, 1982), 342.

35. Editorial, "Clinical Endocrinology," *Endocrinology* 1 (1917): 280–81; Editorial, "The Study of the Internal Secretions: An Introduction," *Endocrinology* 1 (1917): 1–2; Editorial, "The Growing Interest in Endocrinology, *Endocrinology* 1 (1917): 129–30.

36. R. G. Hoskins, *The Tides of Life* (New York: W. W. Norton, 1933), 23–24; Louis Berman, *New Creations in Human Beings* (New York: Doubleday, Dovan, 1938), 301.

37. See Bernard, *Experimental Medicine* (just as Bernard had averred: "It is always a question of separating or altering certain parts of the living machine so as to study them and thus to decide how they function and for what)," 104.

38. Harvey Cushing, "The Hypophysis Cerebri," *Journal of the American Medical Association* [*JAMA*] 53 (1909): 250.

39. Herbert Evans and Joseph Long, "The Effect of the Anterior Lobe Administered Intraperitoneally Upon Growth . . . ," *Anatomical Record* 21 (1921): 62–63.

40. Quoted in Emil Novak, "An Appraisal of Ovarian Therapy," *Endocrinology* 6 (1922): 617.

41. J. M. Anders, "Diagnosis of Myxedema," *American Journal of the Medical Sciences* 160 (1920): 802.

42. National Research Council, "An Appraisal of Endocrinology: A Report Made to the Directors of John and Mary R. Markle Foundation" (Washington, D.C.: National Research Council Press, 1936), 10–11.

43. National Research Council, "An Appraisal of Endocrinology," 11.

44. National Research Council, "An Appraisal of Endocrinology," 20.

45. Henry Christian, "The Use and Abuse of Endocrinology," *The Canadian Medical Association Journal* 14 (1923): 102–6; quotation, 104. With diabetes, "the therapeutic results are so remarkable," declared Henry Christian, the physician-in-chief at Boston's Peter Bent Brigham Hospital in 1923, "that experiments should continue and that new preparations be tested out."

46. Alva Johnson, "Sideshow People," *The New Yorker* (April 14, 1934; April 21, 1934; April 28, 1934), quotations from April 28, 1934, article, 91, 94, 96. The idea of the freak as patient and therefore the freak show as somewhat obscene goes back to the very earliest public acquaintance with endocrinology. See the *Nation* article "Amusement at the Abnormal" (March 19, 1908), which protests "making public sport of what was merely pathological." The case in point was the giant, once considered a superman, and as such worth going to see. But now we understand that he is "the victim of a disease which in other forms kills after horrible disfigurement, and that something at the base of his brain is responsible for the extraordinary and disproportionate growth," 254.

47. J. Mace Andress, A. K. Aldinger, and I. H. Goldberger, *Health Essentials* (Boston: Ginn, 1928), 152–53. The message was repeated in Truman Moon, *Biology for Beginners* (New York: Henry Holt, 1926), 453, again focusing on giants and dwarfs.

48. Dorothy Baruch and Oscar Reiss, *My Body and How It Works*, 4th ed. (New York: Harper and Brothers, 1934), 89.

49. George Bush, Allan Dickie et al., *A Biology of Familiar Things* (New York: American Book, 1939), 91.

50. H. G. Wells, *The Food of the Gods and How It Came to Earth* (New York: Charles Scribner's Sons, 1904), 625.

51. Wells, *The Food of the Gods*, 674.

52. Wells, *The Food of the Gods*, 715.

53. Wells, *The Food of the Gods*, 688–89.

54. Wells, *The Food of the Gods*, 766.

TWO　　The Female Principle

1. There are several excellent historical studies on changing medical under-standing of the female body. See particularly Thomas Laqueur, *Making Sex: Body and Gender from the Greeks to Freud* (Cambridge: Harvard University Press, 1990); Mary Poovey, " 'Scenes of an Indelicate Character': The Medical 'Treatment' of Victorian Women," *Representations* 14 (Spring 1986): 137–68; and Londa Schiebinger, *The Mind Has No Sex?: Women in the Origins of Modern Science* (Cambridge: Harvard University Press, 1989).

2. Charles Birnberg, *Female Sex Endocrinology: Concise Therapy* (Philadel-phia: J. B. Lippincott, 1949), 15.

3. Quoted in Laqueur, *Making Sex*, 149. Quoted in Carroll Smith-Rosenberg and Charles Rosenberg, "The Female Animal: Medical and Biological Views of Women and Her Role in Nineteenth-Century America," *Journal of American History* 60 (1973): 335.

4. Kathy Davis, "Embody-ing Theory: Beyond Modernist and Postmod-ernist Readings of the Body," in Kathy Davis, ed., *Embodied Practices: Feminist Perspectives on the Body* (London: Sage Publications, 1997), 2.

5. George Corner, *The Hormones in Human Reproduction* (Princeton: Prince-ton University Press, 1942), 33–34.

6. Regina Morantz-Sanchez, *Conduct Unbecoming a Woman: Medicine on Trial in Turn-of-the-Century Brooklyn* (New York: Oxford University Press, 1999), 99–113; Ornella Moscucci, *The Science of Woman: Gynaecology and Gender in England, 1800–1929* (Cambridge: Cambridge University Press, 1990), 135–52.

7. Victor Medvei, *A History of Endocrinology* (Hingham, Mass.: MTP Press, 1982), 215.

8. This approach is presented in the seminal monograph, F. H. A. Marshall, *The Physiology of Reproduction* (London: Longmans, Green, 1910), 279.

9. Corner, *Hormones in Human Reproduction*, 80–81.

10. Robert Frank, *The Female Sex Hormone* (Springfield, Ill:. Charles C. Thomas, 1929). In this important study, Frank set out the idea of a "female sex principle." According to Frank, the substance could be found in both the veg-etable and animal kingdoms and could be isolated and measured.

11. Marshall, *Physiology of Reproduction*, 1–2.

12. Frank Lillie, "The Free-Martin: A Study of the Action of Sex Hormones in the Foetal Life of Cattle," *Journal of Experimental Zoology* 23 (1917): 371, 415.

13. For a detailed study of these attitudes and practices, see A. McGehee Har-vey, *Science at the Bedside: Clinical Research in American Medicine, 1905–1945* (Baltimore: Johns Hopkins University Press, 1981).

14. Harvey Cushing, *The Life of Sir William Osler*, vol. 1 (Oxford: Claren-don Press, 1925), 225.

15. Quoted in Medvei, *A History of Endocrinology*, 364.

16. Medvei, *A History of Endocrinology*, 364.

17. George Corner, "The Early History of Oestrogenic Hormones," *Journal of Endocrinology* 31 (1965): vi.

18. Moscucci, *Science of Woman*, 102.

19. Moscucci, *Science of Woman*, 158.

20. Henry Andrews, "The Internal Secretion of the Ovary," *Journal of Obstetrics and Gynecology of the British Empire* 5 (1904): 452.

21. Andrews, "Internal Secretion," 454.

22. Swale Vincent, *Internal Secretion and the Ductless Glands* (New York: Longmans, Green, 1912), 88.

23. Harold Batty Shaw, *Organotherapy in General Practice* (New York: E. W. Carmrick, 1924).

24. Paul Pittenger, "Endocrine Products," *Journal of the American Pharmaceutical Association* 14 (1925): 99–107.

25. For a discussion of the comparative efficacy of these preparations, see: Marshall, *Physiology of Reproduction*, 326–28, and Emil Novak, "Organotherapy of Menstrual Disorders," *Medicine and Surgery* 1 (1917): 576–85.

26. Harry Benjamin, "Eugen Steinach, 1861–1944: A Life of Research," *Scientific Monthly* 61 (1949): 439.

27. Lewellys Barker, "The Principles Underlying Organotherapy and Hormonotherapy," *Endocrinology* 6 (1922): 591–95; R. G. Hoskins, "Some Recent Work on Internal Secretions," *Endocrinology* 6 (1922): 621–32; Emil Novak, *Menstruation and Its Disorders* (New York: D. Appleton, 1921), 314–26.

28. Sophie Aberle and George Corner, *Twenty-five Years of Sex Research* (Philadelphia: W. B. Saunders, 1953), 31.

29. Emil Novak, "An Appraisal of Ovarian Therapy," *Endocrinology* 6 (1922): 617.

30. Novak, "Appraisal of Ovarian Therapy," 617.

31. Novak, "Appraisal of Ovarian Therapy," 618.

32. Two insightful approaches to these issues can be found in Nelly Oudshoorn, *Beyond the Natural Body: An Archaeology of Sex Hormones* (London: Routledge, 1994); and Adele Clarke, "Money, Sex, and Legitimacy at Chicago, Circa 1892–1940: Lillie's Center of Reproductive Biology," *Perspectives on Science* 1 (1993): 367–415.

33. E. C. Dodds, "The New Estrogens," *Edinburgh Medical Journal* 48 (1941): 2–3. For more details, see Charles Mann and Mark Plummer, *The Aspirin Wars: Money, Medicine, and 100 Years of Rampant Competition* (Boston: Harvard Business School Press, 1991), 3–31.

34. Adele Clarke, "Research Materials and Reproductive Science in the United States, 1910–1940," in Gerald Geison, ed., *Physiology in the American Context 1850–1940* (Bethesda, Md.: American Physiological Society, 1987), 323–50.

35. Lillie, "The Free-Martin," 378–79.

36. Florence Sabin, *Franklin Paine Mall* (Baltimore: Johns Hopkins Press, 1934), 213.

37. George Corner, *Anatomist at Large: An Autobiography and Selected Essays* (New York: Basic Books, 1958), 33–34.

38. George Corner, *The Seven Ages of a Medical Scientist: An Autobiography* (Philadelphia: University of Pennsylvania Press, 1981), 159.

39. Alan Parkes, *Off-beat Biologist* (Cambridge, England: Galton Foundation, 1985), 127–28.

40. Edward Doisy, "Isolation of a Crystalline Estrogen from Urine and the Follicular Hormone from Ovaries," *American Journal of Obstetrics and Gynecology* 114 (1972): 701–2.

41. George Corner, "The Early History of Oestrogenic Hormones," *Journal of Endocrinology* 31 (1964–1965): xiii.

42. Edgar Allen and Edward Doisy, "An Ovarian Hormone: Preliminary Report of its Localization, Extraction and Partial Purification and Action in Test Animals," *JAMA* 81 (1923): 819–21.

43. Alan Parkes, "The Rise of Reproductive Endocrinology, 1926–1940," *Journal of Endocrinology* 34 (1966): xxi; Corner, *Hormones in Human Reproduction*, 83–84.

44. Michael Finkelstein, "Professor Bernhard Zondek: An Interview," *Journal of Reproduction and Fertility* 12 (1966): 3–19.

45. Corner, *Hormones in Human Reproduction*, 86.

46. Doisy, "Isolation of a Crystalline Estrogen," 701.

47. Bernhard Zondek, "Estrogenic Hormone in the Urine of the Stallion," *Nature* 133 (1934): 494.

48. Herbert Evans, "Endocrine Glands: Gonads, Pituitary, and Adrenals," *Annual Review of Physiology* 1 (1939): 577.

49. The Laqueur quote can be found in Oudshoorn, *Beyond the Natural Body*, 89.

50. Emil Novak, *Menstruation and Its Disorders* (New York: D. Appleton, 1921), 317.

51. Emil Novak, *Gynecology and Female Endocrinology* (Boston: Little, Brown, 1941), 447.

52. E. C. Dodds, "The New Estrogens," 6–12; Raphael Kurzrok, *The Endocrines in Obstetrics and Gynecology* (Baltimore: Williams & Wilkins, 1937), 52–59; Kenneth Thompson, "Ethinyl Estradiol," *Endocrine Review* 8 (1948): 1088–95. For an overview on products and costs, see Emil Novak, "The Therapeutic Use of Estrogenic Substances," *Glandular Physiology and Therapy: A Symposium* (Chicago: American Medical Association, 1935), 193–210.

53. Laman Gray, "Clinical Study of a New Type of Estrogenic Preparation for Oral Use," *Journal of Clinical Endocrinology* 3 (1943): 92–94; S. Glass and Gordon Rosenblum, "Therapy of the Menopause, Superiority of Conjugated Estrogens—Equine over Diethylstilbestrol," *Journal of Clinical Endocrinology* 3 (1943): 95–97.

54. Elmer Sevringhaus, *Endocrine Therapy in General Practice* (Chicago: Year Book, 1938), 153–59.

55. Novak, *Gynecology,* 447.

56. E. C. Hamblen, *Endocrine Gynecology* (Springfield, Ill.: Charles C. Thomas, 1939), 3. Hamblen goes on to note that "in no branch of medicine does the endocrine system play such a dominant and definitive role as in gynecology" (3).

57. For an extended discussion, see R. G. Hoskins, *Endocrinology: The Glands and Their Functions* (New York: W. W. Norton, 1941).

58. Hamblen, *Endocrine Gynecology,* 93.

59. Hamblen, *Endocrine Gynecology,* 90–91.

60. Charles Mazer and S. Leon Israel, *Diagnosis and Treatment of Menstrual Disorders and Sterility* (New York: Paul B. Hoeber, 1941), 37.

61. Mazer and Israel, *Diagnosis and Treatment,* 38.

62. Jacob Hoffman, *Female Endocrinology* (Philadelphia: W. B. Saunders, 1944), 297–317; Mazer and Israel, *Diagnosis and Treatment,* 140–93.

63. Mazer and Israel, *Diagnosis and Treatment,* 54.

64. Raphael Kurzrok, *The Endocrines,* 159.

65. Emil Novak, *Textbook of Gynecology* (Baltimore: Williams & Wilkins, 1944), 56.

66. Robert Frank, "The Hormonal Causes of Premenstrual Tension," *Archives of Neurology and Psychiatry* 26 (1931): 1053–57.

67. S. Leon Israel, "Premenstrual Tension," *JAMA* 110 (1938): 1721.

68. Israel, "Premenstrual Tension," 1721.

69. Frank, "Hormonal Causes," 1054.

70. Israel, "Premenstrual Tension," 1722.

71. Novak, *Textbook of Gynecology,* 508–10.

72. Hoskins, *Endocrinology,* 300.

73. Hamblen, *Endocrine Gynecology,* 396.

74. Council on Pharmacy and Chemistry, "Estrogenic Substances: Theelin," *JAMA* 100 (1933): 1338.

75. Council on Pharmacy and Chemistry, "Stilbestrol: Preliminary Report of the Council," *JAMA* 113 (1939): 2312.

76. Council on Pharmacy and Chemistry, "Preliminary Report on Progesterone and the Status of Corpus Luteum Hormones Therapy," *JAMA* 116 (1941): 1525.

77. Emil Novak, "The Therapeutic Use of Estrogenic Substances," *JAMA* 104 (1935): 1816.

78. Frank, "Hormonal Causes," 1057.

79. Emil Novak, "The Menopause and Its Management," *JAMA* 110 (1938): 622.

80. Novak, *Gynecology and Female Endocrinology,* 410.

81. Corner, *Hormones*, 99.

82. Novak, *Gynecology*, 450.

83. Novak, "The Therapeutic Use," 1817.

84. Novak, "The Therapeutic Use," 1817.

85. Novak, "The Therapeutic Use," 1818.

86. Novak, "The Therapeutic Use," 1821.

87. Kurzrok, *The Endocrines*, 303.

88. Kurzrok, *The Endocrines*, 279.

89. Editorial, "Contraindications to Estrogen Therapy," *JAMA* 114 (1940): 1560. On the association of malignant growths with estrogen in rodents, see Isabella Perry and Leonard Ginzton, "The Development of Tumors in Female Mice Treated with 1:2: 5:6-Dibenzanthracene and Theelin," *American Journal of Cancer* 29 (1937): 680–703.

90. C. L. Buxton and Earl Engle, "Effects of the Therapeutic Use of Diethylstilbestrol," *JAMA* 113 (1939): 2320.

91. Novak, "The Therapeutic Use," 1821.

92. Novak, *Textbook of Gynecology*, 521.

93. Hoffman, *Female Endocrinology*, 39–41.

94. Elmer Sevringhaus, "Uses and Limitations of Female Sex Endocrine Therapy," in Edward Doisy et al., *Female Sex Hormones* (Philadelphia: University of Pennsylvania Press, 1941), 47.

95. Russell Cecil, *A Textbook of Medicine*, 5th ed. (Philadelphia: W. B. Saunders, 1940), 1376.

96. Hamblen, *Endocrine Gynecology*, 189.

97. Novak, *Textbook of Gynecology*, 536.

98. Hoskins, *Endocrinology*, 297.

99. Hamblen, *Endocrine Gynecology*, 190.

THREE   Making Up, Making Over

1. For a study of drug company advertising in the pursuit of vitamin sales, see Rima Apple, *Vitamania* (New Brunswick, N.J.: Rutgers University Press, 1996).

2. John P. Swann, "The History of Pharmacy and Estates of Science: Institutional Frameworks for Drug Research in America," *Pharmacy in History* 32 (1990): 4.

3. Swann, "The History of Pharmacy," 3–9.

4. The argument here follows closely on Peter Temin, *Taking Your Medicine: Drug Regulation in the United States* (Cambridge: Harvard University Press, 1980), 41–50; Harry Marks, *The Progress of Experiment: Science and the Reform of Therapeutics in the United States, 1900–1990* (New York: Cambridge University Press, 1997).

5. Christopher Kobrak, *National Cultures and International Competition: The Experience of Schering AG, 1851–1950* (Cambridge and New York: Cambridge University Press, 2002); see the introduction and ch. 5, especially 119–20.

6. James H. Madison, *Eli Lilly: A Life, 1885–1977* (Indianapolis: Indiana Historical Society, 1989), 62.

7. C. Rufus Rorem and Robert P. Fischelis, *The Costs of Medicines* Publications of the Committee on the Costs of Medical Care (Chicago: University of Chicago Press, 1932), no. 14, 5–6.

8. Rorem and Fischelis, *The Costs of Medicines,* no. 14, 32.

9. John H. Speer to Gregory Pincus, May 20, 1941, Pincus Mss., Library of Congress, Washington, D.C.

10. R. G. Hoskins, "Some Recent Work on Internal Secretions," *Endocrinology* 6 (1922): 626.

11. Emil Novak, "An Appraisal of Ovarian Therapy," *Endocrinology* 6 (1922): 612, 619.

12. Rand Study, 1940, quoted in Theodore Caplow, "Market Attitudes: A Research Report from the Medical Field," *Harvard Business Review* 30 (1952): 108.

13. Caplow, "Market Attitudes," 105–12. Although the research was conducted a decade later, its relevance to the early 1940s is very high. The sample of physicians had a median age of forty-seven with an average practice history of eighteen years.

14. Ben Gaffin and Associates, "Attitudes of U.S. Physicians Toward the American Pharmaceutical Industry," Opinion Study Financed by the American Medical Association, 1959, American Medical Association Archives, Chicago.

15. Paul C. Olsen, *The Marketing of Drug Products* (New York: Harper & Brothers, 1940), 41.

16. Herbert Menzel and Elihu Katz, "Social Relations and Innovation in the Medical Profession," *Public Opinion Quarterly* 19 (1955–56): 337–52.

17. Caplow, "Market Attitudes," 108–9.

18. See also "Pharmaceutical Advertising Club, Pharmaceutical Advertising," prepared by the Institute for Motivation Research, 1955. Quoted in William D. Benfer, "The Function of the Ethical Pharmaceutical Company Representative in the Marketing of Prescription Drugs with a Given Medical Community" (Ph.D. diss., University of Delaware, 1964), 7–8.

19. Thomas H. Jones, *Detailing the Physician: Sales Promotion by Personal Contact with the Medical and Allied Professions* (New York: Romaine Pierson, 1940).

20. Jones, *Detailing the Physician,* preface, 33, 37, 45–46, 68–69, 121, 195.

21. "Abbott, Laboratories," *Fortune* 22 (August 1940): 63–66.

22. The materials here and below are drawn from the advertising sections of the journal *Endocrinology,* volumes 21 (1937), 25 (1939), 28 (1941), 30 (1942).

23. A. Scott Berg, *Goldwyn: A Biography* (New York: Alfred A. Knopf, 1989), 91–92.

24. Kathryn Fuller, *At the Picture Show: Small-Town Audiences and the Creation of Movie Fan Culture* (Washington, D.C.: Smithsonian Institution Press, 1996), 151.

25. Berg, *Goldwyn*, 115.

26. "Dr. Steinach Coming to Make Old Young," *New York Times*, February 9, 1922, 28. See also, Gertrude Atherton, *Adventures of a Novelist* (New York: Liveright, 1932), 554–56.

27. Atherton, *Adventures of a Novelist*, 558–59.

28. Harry Benjamin, "The Steinach Method as Applied to Women: Preliminary Report," *New York Medical Journal and Medical Record* 118 (1923): 751–52.

29. Gertrude Atherton, *Black Oxen* (New York: Boni and Liveright, 1923), 5, 6.

30. Atherton, *Black Oxen*, 54–55.

31. Atherton, *Black Oxen*, 246.

32. Atherton, *Black Oxen*, 137–38.

33. Atherton, *Black Oxen*, 138.

34. Atherton, *Black Oxen*, 137, 138.

35. Atherton, *Black Oxen*, 172.

36. Atherton, *Black Oxen*, 143.

37. Atherton, *Black Oxen*, 143.

38. Atherton, *Black Oxen*, 176.

39. Atherton, *Black Oxen*, 139.

40. Atherton, *Black Oxen*, 137, 139.

41. Atherton, *Black Oxen*, 176.

42. " '*Black Oxen*' to Be Removed from Shelves of Library," *Rochester Times-Union*, September 29, 1923, 9.

43. Atherton, *Adventures of a Novelist*, 561.

44. Tom Dardis, *Firebrand: The Life of Horace Liveright* (New York: Random House, 1995), 166–67, 169.

45. Gertrude Atherton, "Science After Fifty," *Pictorial Review*, March 1933, 15.

46. Atherton, *Adventures of a Novelist*, 562.

47. Atherton, "Science After Fifty," 15.

48. All the letters quoted below can be found in the Gertrude Franklin Horn Atherton Collection of Papers (BANC-MSFC-H-C 45) at the Bancroft Library, University of California, Berkeley, California.

49. Noreen Price to Gertrude Atherton, May 11, 1923, box 8.

50. Celia Campbell to Gertrude Atherton, February 18, 1923, box 2.

51. Celia Campbell to Gertrude Atherton, February 24, 1923, box 2.

52. Beulah Wettach to Gertrude Atherton, September 18, 1923, box 12.

53. Muriel Pontan to Gertrude Atherton, September 26, 1923, box 8.

54. Noreen Price to Gertrude Atherton, May 11, 1923, box 8.

55. Irma Hams to Gertrude Atherton, January 14, 1936, box 5.

56. Dorothy Vernon to Gertrude Atherton, September 11, 1923, box 11.

57. Mary White to Gertrude Atherton, April 12, 1923, box 12.

58. Helen Stevens to Gertrude Atherton, April 14, 1923, box 8.

59. Saidee Lyford to Gertrude Atherton, December 29, 1935, box 6.

60. Bruce Kellner, *Carl Van Vechten and the Irreverent Decades* (Norman: University of Oklahoma Press, 1968), 108.

FOUR   Controversies Forever

1. William Masters, "Sex Steroid Influence on the Aging Process," *American Journal of Obstetrics and Gynecology* 74 (1957): 733–46.

2. William Masters, "Rationale of Sex Steroid Replacement in the 'Neutral Gender,' " *American Journal of Geriatrics* 3 (1955): 389.

3. Herbert Kupperman, Ben Wetchler, and Meyer Blatt, "Contemporary Therapy of the Menopausal Syndrome," *JAMA* 171 (1959): 1627–37. "One must give serious consideration to long-range steroid therapy in geriatrics, as so well presented by Masters. . . . We are entirely in accord with Masters' concepts and have confirmed his observations," 1632.

4. William Masters and Willard Allen, "Female Sex Hormone Replacement in the Aged Woman: A Preliminary Report," *Journal of Gerontology* 3 (1948): 183.

5. Masters and Allen, "Female Sex Hormone Replacement," 185.

6. Masters and Allen, "Female Sex Hormone Replacement," 183.

7. Masters and Allen, "Female Sex Hormone Replacement," 188.

8. Masters, "Rationale of Sex Steroid Replacement," 389.

9. Masters, "Rationale of Sex Steroid Replacement," 394.

10. This argument is found in Masters, "Rationale of Sex Steroid Replacement," 389–95.

11. Masters, "Rationale of Sex Steroid Replacement," 390.

12. Masters, "Sex Steroid Influence," 733.

13. Masters, "Sex Steroid Influence," 735.

14. Masters, "Sex Steroid Influence," 735.

15. The literature here is a large and interesting one. See, particularly, Fuller Albright, Patricia Smith et al., "Postmenopausal Osteoporosis: Its Clinical Features," *JAMA* 116 (1941): 2465–74; Stanley Wallach and Philip Hennemen, "Prolonged Estrogen Therapy in Postmenopausal Women," *JAMA* 171 (1959): 1637–1642.

16. Masters, "Sex Steroid Influence," 736.

17. See Gregory Pincus, Ralph Dorfman et al., "Steroid Metabolism in Aging

Men and Women," in Gregory Pincus, ed., *Recent Progress in Hormone Research*, vol. 2 (New York: Academic Press, 1955), 307.

18. Masters, "Sex Steroid Influence," 737.

19. Masters, "Sex Steroid Influence," 737–38.

20. Masters, "Sex Steroid Influence," 738.

21. Masters, "Sex Steroid Replacement," 393.

22. Masters, "Sex Steroid Influence," 740.

23. Masters, "Sex Steroid Influence," 744–45.

24. Thomas Mack et al., "Estrogens and Endometrial Cancer in a Retirement Community," *New England Journal of Medicine* (*NEJM*) 294 (1976): 1267; Dianne Kennedy and Carlene Baum, "Noncontraceptive Estrogens: Use Patterns over Time," *Obstetrics and Gynecology* 65 (1985): 441–46.

25. *Physicians' Desk Reference to Pharmaceutical Specialties and Biologicals*, 22nd ed. (Rutherford, N.J.: Medical Economics Co., 1968), 1559; although the language in the *PDR* must meet FDA standards, the volume is given free to doctors by the drug companies. As Peter Temin notes: "Written and financed by the contributing companies and given to doctors, it is a kind of group advertisement." Peter Temin, *Taking Your Medicine: Drug Regulation in the United States* (Cambridge: Harvard University Press, 1980), 89. The advertisement can be found in the *American Journal of Obstetrics and Gynecology* 6 (1966): 3.

26. Ann Walsh, "Pills to Keep Women Young: Eight Distinguished Doctors Who Have Done Important Work with Estrogen Therapy Tell of Its Exciting Results," *McCall's* 93 (October 1965), 105–9.

27. Sherwin Kaufman, "The Truth About Female Hormones," *Ladies' Home Journal* 82 (January 1965): 22–23.

28. Robert A. Wilson, *Feminine Forever* (New York: M. Evans, 1966), 180–81, 15.

29. Nancy Sommers and James Ridgeway, "Can a Woman Be Feminine Forever?," *New Republic*, March 29, 1966: 15–16.

30. Barbara Seaman and Gideon Seaman, *Women and the Crisis in Sex Hormones* (New York: Rawson Associates, 1977), 288–89.

31. Sommers and Ridgeway, "Can a Woman Be Feminine Forever?," 15.

32. Seaman and Seaman, *Crisis in Sex Hormones*, 288.

33. Wilson, *Feminine Forever*, 113–14.

34. Wilson, *Feminine Forever*, 54.

35. Wilson, *Feminine Forever*, 40, 42.

36. Wilson, *Feminine Forever*, 97.

37. Wilson, *Feminine Forever*, 25.

38. Wilson, *Feminine Forever*, 53, 43.

39. Wilson, *Feminine Forever*, 97.

40. Wilson, *Feminine Forever*, 40, 25.

41. Wilson, *Feminine Forever*, 64.

42. Wilson, *Feminine Forever*, 66.

43. *Physicians' Desk Reference*, 1968, 1559.

44. J. Robert Wilson et al., *Obstetrics and Gynecology*, 2nd ed. (St. Louis: C. V. Mosby, 1963), 677.

45. Edmund Novak, *Novak's Textbook of Gynecology*, 7th ed. (Baltimore: Williams & Wilkins, 1965), 646.

46. Helen Jern, *Hormone Therapy of the Menopause and Aging* (Springfield, Ill.: Charles C. Thomas, 1973), 4.

47. Jern, *Hormone Therapy*, 58.

48. Wilson, *Feminine Forever*, 125–27.

49. Wilson, *Feminine Forever*, 128.

50. Wilson, *Feminine Forever*, 128–30.

51. Wilson, *Obstetrics and Gynecology*, 677.

52. Jern, *Hormone Therapy*, 92.

53. The advertisements can be found in *American Journal of Obstetrics and Gynecology* 6 (1966): 81–86.

54. Bill Davidson, "Menopause: Is There a Cure?," *Saturday Evening Post* 240 (1967): 71.

55. Wilson, *Feminine Forever*, 114.

56. Jern, *Hormone Therapy*, 23–24.

57. Jern, *Hormone Therapy*, 26–27.

58. Jern, *Hormone Therapy*, 8–9.

59. Jern, *Hormone Therapy*, 27–28.

60. Edmund Novak, "Replacement Therapy of the Menopause," *Johns Hopkins Medical Journal* 120 (1967): 410, 414.

61. Novak, "Replacement Therapy," 413.

62. Novak, "Replacement Therapy," 410.

63. Novak, "Replacement Therapy," 411.

64. Novak, "Replacement Therapy," 412.

65. Novak, "Replacement Therapy," 410.

66. Novak, "Replacement Therapy," 414.

67. Isabella Perry and Leonard Ginzton, "The Development of Tumors in Female Mice When Treated with 1:2:5:6 Dibenzanthracene and Theelin," *American Journal of Cancer* 29 (1937): 680–704.

68. Articles with similar findings and concerns continued to periodically appear during the 1940s and 1950s. See, for example, William Meissner and Sheldon Sommers, "Endometrial Hyperplasia, Endometrial Carcinoma and Endometriosis Produced Experimentally by Estrogen," *Cancer* 10 (1957): 500–9.

69. Wilson, *Feminine Forever*, 67.

70. See, for example, *Physicians' Desk Reference*, 20th ed. (1966), 543.

71. Jern, *Hormone Therapy*, 150.

72. Jern, *Hormone Therapy*, 155.

73. Jern, *Hormone Therapy*, 76.

74. Sherwin Kaufman, *The Ageless Woman: Menopause, Hormones, and the Quest for Youth* (Englewood Cliffs, N.J.: Prentice Hall, 1967), 46.

75. Novak, "Replacement Therapy," 412.

76. Novak, "Replacement Therapy," 411.

77. Edmund Novak, *Novak's Textbook of Gynecology*, 9th ed. (Baltimore: Williams & Wilkins, 1975), 717–18.

78. Arthur L. Herbst, Howard Ulfelder et al., "Adenocarcinoma of the Vagina: Association of Maternal Stilbestrol Therapy with Tumor Appearance in Young Women," *NEJM* 284 (1971): 878–80.

79. Peter Greenwald, Joseph Barlow et al., "Vaginal Cancer After Maternal Treatment with Synthetic Estrogens," *NEJM* 285 (1971): 390–92.

80. Seaman and Seaman, *Crisis in Sex Hormones*, 13.

81. Harry Ziel and William Finkle, "Increased Risk of Endometrial Carcinoma Among Users of Conjugated Estrogens," *NEJM* 293 (1975): 1167–70.

82. Donald Smith, Ross Prentice et al., "Association of Exogenous Estrogen and Endometrial Carcinoma, *NEJM* 293 (1975): 1164–67.

83. Kenneth Ryan, "Cancer Risk and Estrogen Use in the Menopause," *NEJM* 293 (1975): 1199–1200.

84. Noel Weiss, "Risks and Benefits of Estrogen Use," *NEJM* 293 (1975): 1200–1.

85. Francis McCrea and Gerald Markle, "The Estrogen Replacement Controversy in the USA and UK: Different Answers to the Same Question?," *Social Studies of Science* 14 (1984): 1–26.

86. Seaman and Seaman, *Crisis in Sex Hormones*, 279.

87. Cynthia Cooke and Susan Dworkin, *The Ms. Guide to a Woman's Health* (New York: Berkeley Books, 1981), 310.

88. "FDA Acts on Estrogen Progestin Labeling," *FDA Consumer* (November 1977): 3.

89. "Informing Patients About Estrogens," *FDA Consumer* (November 1976): 8–9.

90. For an excellent overview of the controversy, see Elizabeth Siegel Watkins, " 'Doctor, Are You Trying to Kill Me?': Ambivalence About the Patient Package Insert for Estrogen," *Bulletin of the History of Medicine* 76 (2002): 84–104.

91. Richard Landau, "What You Should Know About Estrogens: Or the Perils of Pauline," *JAMA* 241 (1979): 51.

92. McCrea, "The Politics of Menopause," 11–15; Watkins; " 'Doctor, Are You Trying to Kill Me?,' " 99–100.

93. Elina Hemminki, Diane Kennedy et al., "Prescribing of Noncontraceptive Estrogens and Progestins in the United States 1974–86," *American Journal of Public Health* 78 (1988): 1479–81.

94. Jane Brody, "Menopausal Estrogens: Benefits and Risks of the 'Feminine' Drug," *New York Times,* September 26, 1979, C16.

95. Diane Wysowski, Linda Golden et al., "Use of Menopausal Estrogens and Medroxyprogesterone in the United States, 1982–1992," *Obstetrics and Gynecology* 85 (1995): 6–10.

96. JoAnn Manson and Kathryn Martin, "Postmenopausal Hormone Replacement Therapy," *NEJM* 345 (2001): 34–40.

97. "Estrogen Benefit Is Indicated," *New York Times,* March 7, 1982, 22.

98. Robert Greenblatt and Leland Stoddard, "The Estrogen-Cancer Controversy," *Journal of the American Geriatrics Society* 26 (1978): 8.

99. Anthony Horsman, Mary Jones et al., "The Effect of Estrogen Dose on Postmenopausal Bone Loss," *NEJM* 309 (1983): 1405–7.

100. *Physicians' Desk Reference,* 42nd ed. (Oradell, N.J.: Medical Economics Co., 1988), 665–67.

101. Robert Lindsay, "Estrogen Therapy in the Prevention and Management of Osteoporosis," *American Journal of Obstetrics and Gynecology* 156 (1987): 1347–51.

102. There are still calls for studies to determine the effects of HRT on the risk of hip and other osteoporatic fractures in older women. See Dennis Villareal, Ellen Binder et al., "Bone Mineral Density Response to Estrogen Replacement in Frail Elderly Women," *JAMA* 286 (2001): 815–21.

103. Manson and Martin, "Postmenopausal Hormone Replacement Therapy," 34.

104. Stephen Hulley, Deborah Grady et al., "Randomized Trial of Estrogen Plus Progestin for Secondary Prevention of Coronary Heart Disease in Postmenopausal Women," *JAMA* 280 (1998): 605–13.

105. David Herrington, David Reboussin et al., "Effects of Estrogen Replacement on the Progression of Coronary-Artery Atherosclerosis," *NEJM* 343 (2000): 522–29.

106. Manson and Martin, "Hormone Replacement Therapy," 38.

107. Lori Mosca, Peter Collins et al., "Hormone Replacement Therapy and Cardiovascular Disease: A Statement for Healthcare Professionals from the American Heart Association," *Circulation* 104 (2001): 502.

108. Sally Shumaker, Beth Reboussin et al., "The Women's Health Initiative Memory Study (WHIMS): A Trial of the Effort of Estrogen Therapy in Preventing and Slowing the Progression of Dementia," *Controlled Clinical Trials* 19 (1998): 605.

109. E. Hogervorst, J. Williams et al., "The Nature of the Effect of Female Gonadal Hormone Replacement Therapy on Cognitive Function in Postmenopausal Women: A Meta-Analysis," *Neuroscience* 101 (2001): 487.

110. Ruth Mulnard, Carl Cotman et al., "Estrogen Replacement Therapy for Treatment of Mild to Moderate Alzheimer Disease: A Randomized Controlled Trial," *JAMA* 283 (2000): 1007–8.

111. See Bennett Shaywitz and Sally Shaywitz, "Estrogen and Alzheimer Disease: Plausible Theory, Negative Clinical Trial," *JAMA* 283 (2001): 1055.

112. For a summary, see Hogervorst, "The Effect of . . . Hormone Replacement Therapy," 485–551.

113. Kristine Yaffe, George Sawaya et al., "Estrogen Therapy in Postmenopausal Women: Effects on Cognitive Function and Dementia," *JAMA* 279 (1998): 688–95.

114. Erin LeBlanc, Jeri Janowsky et al., "Hormone Replacement Therapy and Cognition: Systematic Review and Meta-analysis," *JAMA* 285 (2001): 1489–99.

115. LeBlanc, "Hormone Replacement Therapy," 1498; Catherine DeAngelis and Margaret Winker, "Women's Health—Filling the Gaps," *JAMA* 285 (2001): 1509.

116. Mulnard, "Estrogen Replacement Therapy," 1013.

117. Mulnard, "Estrogen Replacement Therapy," 1015.

118. Kristine Yaffe, Kathryn Krueger et al., "Cognitive Function in Postmenopausal Women," *NEJM* 344 (2001): 1207–13.

119. Richard Mayeux, "Can Estrogen or Selective Estrogen-Receptor Modulators Preserve Cognitive Function in Elderly Women?," *NEJM* 344 (2001): 1242–44.

120. Jane Cauley, Steven Cummings et al., "Prevalence and Determinants of Estrogen Replacement Therapy in Elderly Women," *American Journal of Obstetrics and Gynecology* 163 (1990): 1438–44.

121. Carol Darby, Anne Hume et al., "Correlates of Postmenopausal Estrogen Use and Trends Through the 1980's in Two Southeastern New England Communities," *American Journal of Epidemiology* 137 (1993): 1125–35.

122. Sally McNagny, Nanette Wenger et al., "Personal Use of Postmenopausal Hormone Replacement Therapy by Women Physicians in the United States," *Annals of Internal Medicine* 27 (1997): 1093–96.

123. Jane Marsh, Kate Brett et al., "Racial Differences in Hormone Replacement Therapy Prescriptions," *Journal of Obstetrics and Gynecology* 93 (1999): 999–1003.

124. Robert Reynolds, Alexander Walker et al., "Discontinuation of Postmenopausal Hormone Therapy in a Massachusetts HMO," *Journal of Clinical Epidemiology* 54 (2001): 1056–64.

125. American College of Obstetricians and Gynecologists, Committee on Gynecologic Practice, "Risk of Breast Cancer with Estrogen-Progestin Therapy," *Journal of Obstetrics and Gynecology* 98 (2001): 1181–83.

126. American Association of Clinical Endocrinologists, Press Release, February 24, 2000, www.aace.com/pub/press/releases/index (December 26, 2001).

127. American College of Obstetricians and Gynecologists, "Risk of Breast Cancer," 1181.

128. American Association of Clinical Endocrinologists, "Medical Guidelines for Clinical Practice for Management of Menopause," *Endocrine Practice* 5 (1999): 357.

129. The Women's Health Initiative Study Group, "Design of the Women's Health Initiative Clinical Trial and Observational Study," *Controlled Clinical Trials* 19 (1998): 61–109.

130. Writing Group for the Women's Health Initiative Investigators, "Risks and Benefits of Estrogen Plus Progestin in Healthy Postmenopausal Women: Principal Results from the Women's Health Initiative Randomized Controlled Trial," *JAMA* 288 (2002): 321–33.

131. Suzanne Fletcher and Graham Colditz, "Failure of Estrogen Plus Progestin Therapy for Prevention," *JAMA* 288 (2002): 366–67.

132. American Heart Association, "Media Advisory," July 9, 2002.

133. Gina Kolata and Melody Petersen, "Hormone Replacement Study a Shock to the Medical System," *New York Times,* July 10, 2002, A1.

134. Kolata and Petersen, "Hormone Replacement Study," A16.

135. American Association of Clinical Endocrinologists, Press Release, July 9, 2002, www.aace.com/pubpress/hrt.php (July 13, 2002).

136. Wulf Utian, "Managing Menopause After HERS II and WHI: Coping with the Aftermath," *Menopause Management* (July/August 2002): 6–7.

137. Ron Winslow and Scott Hensley, "Wyeth Sales Team Calls on Doctors After Study," *Wall Street Journal,* July 17, 2002, D3.

138. Sally Shumaker, Claudine Legault et al., "Estrogen Plus Progestin in the Incidence of Dementia and Mild Cognitive Impairment in Postmenopausal Women," *JAMA* 289 (2003): 2651–72.

139. Rowan Chlebowski, Susan Hendrix et al., "Influence of Estrogen Plus Progestin on Breast Cancer and Mammography in Healthy Postmenopausal Women," *JAMA* 289 (2003): 3248–53.

140. Peter Gann and Monica Morrow, "Combined Hormone Therapy and Breast Cancer: A Single-Edged Sword," *JAMA* 289 (2003): 3304

141. Jennifer Hays, Judith Ockene et al., "Effects of Estrogen plus Progestin on Health-Related Quality of Life," *NEJM* 348 (2003): 1839–54.

142. Editorial, "Delusions of Feeling Better," *New York Times,* March 19, 2003, A28.

143. Roy Moynihan, "Drug Company Secretly Briefed Medical Societies on HRT," *British Medical Journal* 326 (2003): 1161.

144. Kristine Yaffe, "Hormone Therapy and the Brain: Deja Vu All Over Again?" *JAMA* 289 (2003): 2718.

145. See Caren Solomon and Robert Dluhy, "Rethinking Postmenopausal Hormone Therapy," *NEJM* 348 (2003): 579–80; Francine Grodstein, Thomas Clarkson et al., "Understanding the Divergent Data on Postmenopausal Hormone Therapy," *NEJM* 348 (2003): 645–50.

FIVE   The Body As Turf

1. For an excellent history, see Elizabeth Haiken, *Venus Envy: A History of Cosmetic Surgery* (Baltimore: Johns Hopkins University Press, 1997). See also Kathy Peiss, *Hope in a Jar: The Making of America's Beauty Culture* (New York: Henry Holt and Company, 1999).

2. Haiken, *Venus Envy*, 31.

3. Sander Gilman, *Making the Body Beautiful: A Cultural History of Aesthetic Surgery* (Princeton: Princeton University Press, 1999), provides a fascinating account of the attraction of stigmatized groups to plastic surgery. Drawing on medical and cultural historical sources, Gilman discusses the importance of "passing" to the surgeons and their patients.

4. Haiken, *Venus Envy*, 123.

5. Yves-Gerard Illouz, "Surgical Remodeling of the Silhouette by Aspiration Lipolysis or Selective Lipectomy," *Aesthetic Plastic Surgery* 9 (1985): 7–21.

6. Ivo Pitanguy, "Evaluation of Body Contouring Surgery Today: A 30 Year Perspective," *Plastic and Reconstructive Surgery* 105 (2000): 1499–1514.

7. Illouz, "Surgical Remodeling," 7. See also Francis Otteni and Pierre Fournier, "A History and Comparison of Suction Techniques Until Their Debut in North America," in Gregory Hetter, ed., *Lipoplasty: The Theory and Practice of Blunt Suction Lipectomy*, 2nd ed. (Boston: Little, Brown, 1990), 23.

8. Pierre Fournier, "Popularization of the Technique," in Hetter, *Lipoplasty*, 35–36. On the complications in earlier methods, see also R. L. Dolsky, "Body Sculpturing by Lipo-Suction Extraction," *Aesthetic Plastic Surgery* 8 (1984): 75–83.

9. Yves-Gerard Illouz, "The Origins of Lipolysis," in Hetter, *Lipoplasty*, 27–34; Fournier, "Popularization," 35.

10. An article on the history of liposuction by two dermatologists refers to Illouz as a gynecologist, perhaps because of his use of the Karman cannula. See Naomi Lawrence and William Coleman III, "Liposuction," *Advances in Dermatology* 11 (1996): 19.

11. Personal interview, June 1995.

12. Yves-Gerard Illouz and Yves de Villers, *Body Sculpturing by Lipoplasty* (New York: Churchill Livingstone, 1989), 10–11; Illouz, "The Origins of Lipolysis," 27–34.

13. Personal interview, June 1995.

14. For information on Illouz, see illouz.com, March 21, 2002.

15. DeVillers, foreword, in Illouz, *Body Sculpturing*, vi; Illouz, preface, *Body Sculpturing*, vii.

16. Norman Martin, "Introduction of Lipolysis to the United States," in Hetter, *Lipoplasty*, 39.

17. Martin, "Introduction," 38.

18. Gregory Hetter, "Closed Suction Lipoplasty on 1078 Patients: Illouz Told the Truth," *Aesthetic Plastic Surgery* 12 (1988): 183–85.

19. Gregory Hetter, Preface to the First Edition, in Hetter, ed., *Lipoplasty*, 2nd ed., xi.

20. See American Society of Plastic and Reconstructive Surgeons, "The History of the ASPRS," *Journal of Plastic and Reconstructive Surgery* 94 (1994): 79A; and Mark Gorney, "Sucking Fat: An 18-Year Statistical and Personal Retrospective," *Plastic and Reconstructive Surgery* 107 (2001): 608.

21. The Ad Hoc Committee on New Procedures of the American Society of Plastic and Reconstructive Surgeons, "Report of Suction-Assisted Lipectomy," in Illouz, *Body Sculpturing*, 467.

22. Ad Hoc Committee, "Report," 463.

23. Ad Hoc Committee, "Report," 464.

24. Ad Hoc Committee, "Report," 464.

25. Ad Hoc Committee, "Report," 464.

26. Ad Hoc Committee, "Report," 463, 465.

27. Ad Hoc Committee, "Report," 466.

28. Yves-Gerard Illouz, "Body Contouring by Lipolysis: A 5-Year Experience with over 3000 Cases," *Plastic and Reconstructive Surgery* 72 (1983): 591–97. Illouz reported 81 percent of his 1,326 patients underwent multiple procedures.

29. Ad Hoc Committee, "Report," 464.

30. Ad Hoc Committee, "Report," 465.

31. Claire Fox and William Graham III, "The American Board of Plastic Surgery 1937–1987," *Plastic and Reconstructive Surgery* 82 (1988): 166–85.

32. Thomas Baker, "The American Society for Aesthetic Plastic Surgery: A Look at Its Origin, Status, and Future," *Plastic and Reconstructive Surgery* 70 (1982): 602.

33. Frederick Grazer and Fred Meister, "Complications of the Tumescent Formula for Liposuction," *Plastic and Reconstructive Surgery* 100 (1997): 1894.

34. Lawrence and Coleman, "Liposuction," 20.

35. Joseph McCarthy (ed.), *Plastic Surgery: General Principles*, vol. 1 (Philadelphia: W. B. Saunders, 1990), 114.

36. Hetter, "Blood and Fluid Replacements," in Hetter, *Lipoplasty*, 191.

37. American Society of Plastic and Reconstructive Surgeons, "The History of the ASPRS," *Journal of Plastic and Reconstructive Surgery* 94 (1994): 79A.

38. Illouz, *Body Sculpturing*, 13–14.

39. Julius Newman, "Lipo-Suction Surgery: Past-Present-Future," *American Journal of Cosmetic Surgery* 1 (1984): 20.

40. Mark Sheehan, "What the *Goldfarb* Decision Means to the Medical Profession," *Journal of Legal Medicine* (November–December 1975): 21–29.

41. American Society of Plastic and Reconstructive Surgeons, "The History of the ASPRS," 72A–74A.

42. Stu Chapman, "California's King of Cosmetic Surgery," *Legal Aspects of Medical Practice* (February 1979): 47–48.

43. Mark Gorney, "Plastic Surgery Is Not a Casual Thing," *Annals of Plastic Surgery* 1 (1978): 531.

44. Richard Caimbrone, "Yellow Pages: The Cornerstone for 'Professional' Advertising," *The Implant Society* 5 (1995): 13–15. Deborah Sullivan, *Cosmetic Surgery: The Cutting Edge of Commercial Medicine in America* (New Brunswick, N.J.: Rutgers University Press, 2001), discusses how cosmetic surgery encouraged commercialism in medicine but does not discuss liposuction.

45. Theodore Tromovitch, "Beginning Dermatologic Surgery: An Essay," *Journal of Dermatologic Surgery* 1 (1975): 63–64.

46. Tromovitch, "Beginning Dermatologic Surgery," 64.

47. Edward Krull, "Reflections on Skin Surgery," *Journal of Dermatologic Surgery* 2 (1976): 400.

48. Theodore Tromovitch, Samuel Stegman et al., "A Survey of Dermatologic Surgery Procedures," *Journal of Dermatologic Surgery and Oncology* 13 (1987): 764.

49. Bruce Chrisman and William Coleman III, "Determining Safe Limits for Untransfused, Outpatient Liposuction: Personal Experience and Review of the Literature," *Journal of Dermatologic Surgery and Oncology* 10 (1988): 1097.

50. Samuel Stegman, "The Application of Lipo-Suction Surgery in Dermatology," in Jeffrey Callen, ed., *Advances in Dermatology*, vol. 1 (Chicago: Year Book Medical Publishers, 1986), 213.

51. Jeffrey Klein, "The Tumescent Technique for Lipo-Suction Surgery," *Journal of Cosmetic Surgery* 4 (1987): 263.

52. Jeffrey Klein, "Tumescent Technique for Regional Anesthesia Permits Lidocaine Doses of 35mg/kg for Liposuction," *Journal of Dermatologic Surgery and Oncology* 16 (1990): 249.

53. Jeffrey Klein, "Tumescent Technique Chronicles: Local Anesthesia, Liposuction and Beyond," *Dermatologic Surgery* 21 (1995): 452.

54. Klein, "The Tumescent Technique for Lipo-Suction Surgery," 266.

55. Klein, "Tumescent Technique Chronicles," 452.

56. Klein, "Tumescent Technique for Regional Anesthesia," 248–62.

57. Gerald Pitman, John Aker et al., "Tumescent Liposuction: A Surgeon's Perspective," *Clinics in Plastic Surgery* 4 (1996): 640.

58. Scott Replogle, "The 'Standard Technique' of Liposuction: Viewpoint from a Plastic Surgeon," *Dermatologic Clinics* 8 (1990): 455.

59. Replogle, "The 'Standard Technique,'" 454–55.

60. American Academy of Dermatology, "Guidelines of Care for Liposuction," *Journal of the American Academy of Dermatology* 24 (1991): 489–94.

61. Ariel Ostad, Nobu Kageyama et al., "Tumescent Anesthesia with a Lido-

caine Dose of 55 mg/kg Is Safe for Liposuction," *Dermatologic Surgery* 22 (1996): 921–27.

62. Mark Gilliland, George Commons et al., "Safety Issues in Ultrasound-Assisted Large Volume Lipoplasty," *Clinics in Plastic Surgery* 26 (1999): 318.

63. Rudolph de Jong and Frederick Grazer, "Titanic Tumescent Anesthesia," *Dermatologic Surgery* 24 (1998): 689–90.

64. Kenneth Shestak, "Marriage Abdominoplasty Expands the Mini-Abdominoplasty Concept," *Plastic and Reconstructive Surgery* 103 (1999): 1020–31.

65. Daniel Morello, Gustavo Colon et al., "Patient Safety in Accredited Office Surgical Facilities," *Plastic and Reconstructive Surgery* 99 (1997): 1496.

66. John Penn and James Baker, "The Office-Based Elective Surgery Center," *Annals of Plastic Surgery* 4 (1980): 94–99.

67. Debra Weissman, "Contending for the Future," Lecture to Division of Plastic Surgery, Columbia Presbyterian Medical Center, New York, New York, April 22, 1996. See also Debra Weissman, "Is It Time for a Checkup of Your Front Desk?," *Journal of Medical Practice Management* 11 (1995): 121–22.

68. Gustavo Colon and Robert Singer, "Introduction of New Technology to the Office," *Clinics in Plastic Surgery* 26 (1999): 356.

69. Colon and Singer, "Introduction of New Technology," 358.

70. Jane Haher, "Ancillary Aesthetic Procedures: 'Finishing Touches,' " Lecture at St. Vincent's Hospital, March 21, 1996.

71. "Lunchtime Cosmetic Surgery: Expert Cosmetic Surgeon Reveals New Secrets for Quick-Fix Rejuvenation," *Business Wire*, February 25, 1998.

72. Robert Goldwyn, *Beyond Appearance: Reflections of a Plastic Surgeon* (New York: Dodd, Mead, 1986), 124–25.

73. Aaron Elstein, "Some Sub-Prime Auto Lenders Doing Body Work on the Side," *American Banker*, November 14, 1996, 24.

74. Shira Boss, "Elective Surgery Without the Plastic: Low-Interest Medical Financing Provides Alternative to Credit Cards," *Crain's New York Business* (June 22, 1998): 26.

75. Tolbert Wilkinson, "Radio & TV Marketing," *Technical Forum* (April 23, 1997): 5.

76. Elstein, "Some Sub-Prime Auto Lenders," 24.

77. Boss, "Elective Surgery Without the Plastic," 26.

78. United States Congress, House of Representatives 101st Congress, Committee on Small Business, Subcommittee on Regulation, Business Opportunities and Energy, "Unqualified Doctors Performing Cosmetic Surgery: Policies and Enforcement Activities of the Federal Trade Commission," part 1, April 4, 1989 (Washington, D.C.: Government Printing Office, 1989–1990), 30.

79. Carl Manstein, "What Is Wrong with Advertising," *Plastic and Reconstructive Surgery* 98 (1996): 1124–25.

80. Manstein, "What Is Wrong with Advertising," 1125.

81. Madelyn Schwartz Quattrone, "Is the Physician's Office the Wild, Wild West of Health Care?," *Journal of Ambulatory Care Management* 23 (2000): 66.

82. Replogle, "The 'Standard Technique,' " 453.

83. Grazer and Meister, "Complications of the Tumescent Formula," 1894.

84. Robert Goldwyn, *The Patient and the Plastic Surgeon,* 2nd ed. (Boston: Little, Brown, 1991), 338–39.

85. Gerald Pitman, *Liposuction & Aesthetic Surgery* (St. Louis: Quality Medical Publishing, 1993), 5–7.

86. C. William Hanke, Gerald Bernstein et al., "Safety of Tumescent Liposuction in 15,336 Patients: National Survey Results," *Dermatologic Surgery* 21 (1995): 459.

87. Information on state regulations can be found at aorn.org/journal/2001 (September 20, 2001).

88. Rama Rao, Susan Ely et al., "Deaths Related to Liposuction," *NEJM* 340 (1999): 1471–75.

89. Frederick Grazer and Rudolph de Jong, "Fatal Outcomes from Liposuction: Census Survey of Cosmetic Surgeons," *Plastic and Reconstructive Surgery* 105 (2000): 437–38.

90. Grazer and de Jong, "Fatal Outcomes," 441–42.

91. Grazer and de Jong, "Fatal Outcomes," 444.

92. Quattrone, "The Wild, Wild West of Health Care?," 68.

93. For a detailed account of the efforts of some states to regulate outpatient cosmetic surgery, see Sullivan, *Cosmetic Surgery,* 189–200; and on Florida in particular, see Sue Landry, "Board Eases Office Surgery Limit," *St. Petersburg Times,* June 5, 1999, 1B.

94. Jon Sutton, "Office-Based Surgery Regulation: Improving Patient Safety and Quality Care," *Bulletin of the American College of Surgeons* 86 (2001): 11–12.

95. New York State Department of Health, Committee on Quality Assurance in Office-Based Surgery, *Clinical Guidelines for Office-Based Surgery: A Report to the New York State Public Health Council,* unpublished report (December 2000), 1–10. For details on the court action, see mssny.org (January 24, 2003).

96. Mark Gorney, "Liability in Suction-Assisted Lipoplasty: A Different Perspective," *Clinics in Plastic Surgery* 26 (1999): 445.

97. Simon Fredericks, "Analysis and Introduction of a Technology: Ultrasound-Assisted Lipoplasty Task Force," *Clinics in Plastic Surgery* 26 (1999): 195.

98. plasticsurgery.org (September 1, 2001).

99. William Coleman III, William Hanke et al., "Does the Location of the Surgery or the Specialty of the Physician Affect Malpractice Claims in Liposuction?," *Dermatologic Surgery* 25 (1999): 343–47.

100. American Academy of Dermatology, "Guidelines of Care for Liposuction," *Journal of the American Academy of Dermatology* 45 (2001): 438–47.

101. Institute of Medicine, *To Err Is Human: Building a Safer Health Care System* (Washington, D.C.: National Academy Press, 2000), lists states with mandatory adverse event reporting systems and the facilities in the state that must report adverse events. See appendix D, "Characteristics of State Adverse Event Reporting Systems," 254–65.

102. Institute of Medicine, *To Err Is Human*, 165.

103. We interviewed twenty consumers who had undergone liposuction between 1996 and 1998 in New York City. The consumers were referred to us by plastic surgeons. They represent a sample of convenience; their responses are not presumed to be statistically representative of all liposuction users. We also attended question-and-answer sessions led by New York plastic surgeons who were explaining the procedure to women from diverse ethnic and racial groups.

104. The American Society for Aesthetic Plastic Surgery reported that their members performed more than 385,000 liposuction procedures in 2001, surgery.org (February 18, 2003). Liposuction remained the number one procedure.

105. Through the 1980s, surgeons and consumers thought that cosmetic surgery was appropriate for women, not men. See Diana Dull and Candace West, "Accounting for Cosmetic Surgery: The Accomplishment of Gender," *Social Problems* 38 (1991): 54–70.

106. webmd.com and mayoclinic.com (August 29, 2001).

107. liposite.com (August 30, 2001).

108. Emin Babakus, Steven Remington et al., "Issues in the Practice of Cosmetic Surgery: Consumers' Use of Information and Perceptions of Service Quality," *Journal of Health Care Marketing* 11 (1991): 12–19.

109. Some women who have undergone other forms of cosmetic surgery also insist that they are pleasing only themselves. See Dull and West, "Accounting for Cosmetic Surgery," 61–62.

110. The importance of the mirror has been explored at great length by psychologists. See April Fallon, "Culture in the Mirror: Sociocultural Determinants of Body Image," in Thomas Cash and Thomas Pruzinsky, eds., *Body Images: Development, Deviance, and Change* (New York: Guilford Press, 1990), 80–109.

111. Psychologists report that happiness is a goal of cosmetic surgery and explore why some consumers fail to achieve it. See Thomas Pruzinsky and Milton Edgerton, "Body-Image Change in Cosmetic Surgery," in Cash and Pruzinsky, *Body Images*, 217–36.

112. One survey reported that 76 percent of liposuction users are satisfied with the result. See Eric Dillerud and Lise Haheim, "Long Term Results of Blunt Suction Lipectomy Assessed by a Patient Questionnaire Survey," *Plastic and Reconstructive Surgery* 92 (1993): 35–40.

113. Katharine Phillips, "Body Dysmorphic Disorder: The Distress of Imagined Ugliness," *American Journal of Psychiatry* 148 (1991): 1138–49.

114. Mike Featherstone, "The Body in Consumer Culture," *Theory, Culture and Society* 1 (1982): 21.

115. Kathy Davis, "Embody-ing Theory: Beyond Modernist and Postmodernist Readings of the Body," in Kathy Davis, ed., *Embodied Practices: Feminist Perspectives on the Body* (London: Sage Publications, 1997), 2.

116. Kathy Davis, *Reshaping the Female Body: The Dilemma of Cosmetic Surgery* (New York: Routledge, 1995), 11.

117. Davis, "Embody-ing Theory," 10.

118. Susan Bordo, " 'Material Girl': The Effacements of Postmodern Culture," *Michigan Quarterly Review* 29 (1990): 654, 676.

119. Annette Kuhn, *The Power of the Image: Essays on Representation and Sexuality* (London: Routledge & Kegan Paul, 1985), 13.

120. Jane Gaines, "Fabricating the Female Body," in Jane Gaines and Charlotte Herzog, *Fabrications: Costume and the Female Body* (New York: Routledge, 1990), 6.

121. Quoted in Gaines, "Fabricating the Female Body," 9.

122. Gaines, "Fabricating the Female Body," 9.

SIX   Borrowed Manhood

1. George Corner, *The Hormones in Human Reproduction* (Princeton: Princeton University Press, 1942), 228–29.

2. Charles E. Brown-Séquard, "The Effects Produced on Man by Subcutaneous Injections of a Liquid Obtained from the Testicles of Animals," *Lancet* 2 (1899): 105–7.

3. Charles. E. Brown-Séquard, "On a New Therapeutic Method Consisting in the Use of Organic Liquids Extracted from the Glands and Other Organs," *British Medical Journal* 1 (1893): 1212–14.

4. The story is well told in Max Thorek, *The Human Testis* (Philadelphia: J. B. Lippincott, 1924), 37–40. It is also prominent in Swale Vincent, *Internal Secretion of the Ductless Glands* (New York: Longmans, Green, 1912), 70–71.

5. Swale Vincent, *Internal Secretion and the Ductless Glands* (London: Arnold, 1912), 70–71.

6. Corner, *Hormones in Human Reproduction,* 235.

7. Harry Benjamin, "Eugen Steinach, 1861–1944: A Life of Research," *Scientific Monthly* 61 (1945): 427–42.

8. Eugen Steinach, "Biological Methods Against the Process of Old Age," *Medical Journal and Record* 125 (1927): 78.

9. Steinach, "Biological Methods," 78.

10. Eugen Steinach, *Sex and Life* (New York: Viking Press, 1940), 8–9.

11. Steinach, *Sex and Life,* 8–10.

12. Steinach, *Sex and Life,* 130.

13. Steinach, *Sex and Life,* 78–80.

14. Harry Benjamin, "Preliminary Communication Regarding Steinach's Method of Rejuvenation," *New York Medical Journal* 114 (1921): 687–88.

15. Harry Benjamin, "New Clinical Aspects of the Steinach Operation," *Medical Journal and Record* 122 (1925): 589, 593–94.

16. Harry Benjamin, "Theory and Practice of the Steinach Operation," *New York Medical Journal* 116 (1922): 206.

17. Benjamin, "New Clinical Aspects," 592.

18. Victor Lespinasse, "Transplantation of the Testicle," *JAMA* 61 (1913): 1869–70. A colleague of Lespinasse, Frank Lydston, even hypothesized that cancer tumors might be susceptible to hormonal influences, on the grounds that cancers appeared more frequently with the approach of menopause in women and advanced age in men. Lydston was hopeful that "sex gland hormone therapy" would prove a highly useful cancer treatment.

19. L. L. Stanley, "Testicular Substance Implant," *Endocrinology* 5 (1921): 712.

20. L. L. Stanley, "An Analysis of One Thousand Testicular Substance Implantations," *Endocrinology* 6 (1922): 787–93. There were others as well involved in this venture, including H. Lyons Hunt, "Experiences in Testicle Transplantation," *Endocrinology* 6 (1922): 652–54; and Max Thorek, "The Present Position of Testicle Transplantation in Surgical Practice: A Preliminary Report of a New Method," *Endocrinology* 6 (1922): 771–75.

21. David Hamilton, *The Monkey Gland Affair* (London: Chatto and Windus, 1986).

22. Max Thorek, *A Surgeon's World* (New York: Lippincott, 1943), 170, 185, 191.

23. Max Thorek, "The Male Climacterium, Its Clinical Significance and Treatment," *Medical Journal and Record* 119 (1924), supplement, xlvii.

24. Max Thorek, "Experimental Investigations of the Role of the Leydig, Seminiferous and Sertoli Cells and Effects of Testicular Transplantation," *Endocrinology* 8 (1924): 89; Max Thorek, "The Gonads and the Large Bowel in Dementia Praecox," *Clinical Medicine and Surgery* 34 (1927): 348–53.

25. Max Thorek, *Plastic Surgery of the Breast and Abdominal Wall* (Springfield, Ill.: Charles C. Thomas, 1942), 168.

26. Thorek, *Plastic Surgery of the Breast and Abdominal Wall,* ix, x, xiii, 168.

27. Walter M. Kearns, "The Clinical Application of Testosterone," *JAMA* 112 (1939): 2256.

28. Editorial, "Empiricism," *The Endocrine Survey* 5 (1928): 295.

29. Editorial, "Empiricism," 295.

30. Editorial, "The Rational Attitude Towards Endocrine Therapy," *The Endocrine Survey* 4 (1927): 169.

31. Editorial, "Empiricism in Medicine," *The Endocrine Survey* 2 (1925): 44–45.

32. Editorial, "The Clinical Employment of Endocrines," *The Endocrine Survey* 2 (1925): 47.

33. Editorial, "Endocrinology-Clinical and Experimental," *The Endocrine Survey* 2 (1925): 85.

34. Editorial, "Endocrinology," 86.

35. Editorial, "Rational Therapy," *The Endocrine Survey* 2 (1925): 386.

36. Editorial, "Treating the Patient Individually," *The Endocrine Survey* 2 (1925): 423.

37. Editorial, "Endocrinology," 87.

38. Editorial, " 'Specializing' in Endocrinology," *The Endocrine Survey* 2 (1925): 132–33.

39. Editorial, "Rational Therapy," 387.

40. Editorial, "Practical Organotherapy," *The Endocrine Survey* 2 (1925): 300.

41. Editorial, " 'Things That Doctors Do Not Know,' " *The Endocrine Survey* 3 (1926): 464.

42. Editorial, "Treating the Patient," 423.

43. Editorial, "The Clinical Employment," 48.

44. Evans's quote can be found in Hans Lisser, "The Endocrine Society: The First Forty Years (1917–1957)," *Endocrinology* 80 (1967): 7.

45. Lisser, "The Endocrine Society," 7.

46. Editorial, "Disappointments of Endocrinology," *JAMA* 76 (1921): 1685.

47. "The Present Position of Organotherapy," *Lancet* 104 (1923): 131.

48. Editorial, "The Endocrine Glands—a Caution," *JAMA* 76 (1921): 1500.

49. Lewellys Barker, "The Principles Underlying Organotherapy and Hormonotherapy," *Endocrinology* 6 (1922): 595.

50. Barker, "The Principles Underlying Organotherapy," 591–95.

51. Henry Christian, "The Use and Abuse of Endocrinology," *Canadian Medical Association Journal* 14 (1924): 102–6; Lemuel McGee, "The Effects of the Injection of a Lipoid Fraction of Bull Testicle in Capons," *Proceedings of the Institute of Medicine of Chicago* 5 (1927): 242–51.

52. See Robert Kohler, *From Medical Chemistry to Biochemistry* (Cambridge: Cambridge University Press, 1982).

53. Carl Moore, "The Regulation of Production and the Function of the Male Sex Hormone," *JAMA* 97 (1931): 519.

54. Moore, "The Regulation of Production," 521–22.

55. Adolph Butenandt, "Chemical Constitution of the Follicular and Testicular Hormones," *Nature* 130 (1932): 238.

56. Corner, *The Hormones in Human Reproduction*, 232–33.

57. James Hamilton, "Treatment of Sexual Underdevelopment with Synthetic Male Hormone Substance," *Endocrinology* 21 (1937): 649–54.

58. Samuel Vest, Jr., and John Eager Howard, "Clinical Experiments with the Use of Male Sex Hormones," *Journal of Urology* 40 (1938): 154–83.

59. Walter Kearns, "The Clinical Application of Testosterone," *JAMA* 112 (1939): 2257. See also Hans Lisser and L. E. Curtis, "Testosterone Therapy of Male Eunuchoids," *Journal of Clinical Endocrinology* 2 (1943): 396–97.

60. Robert Escamilla and Hans Lisser, "Testosterone Therapies of Eunuchoids," *Journal of Clinical Endocrinology* 1 (1941): 635.

61. The effectiveness of the compound also appeared in the treatment of young boys with undescended testicles (a not uncommon birth defect). Although the correction could be made through surgery, physicians were looking for a less invasive method. Testosterone, they speculated, would enlarge the testes and thereby increase the likelihood of their descending from the abdominal area to the scrotum. The actual results varied considerably.

62. Henry Turner, "Male Sex Hormone," *Endocrinology* 24 (1939): 773. See also Allan Kenyon, "Problems in the Recognition and Treatment of Testicular Insufficiency," *NEJM* 225 (1941): 714–18.

63. Stanley Goldman and Mark Markham, "Clinical Use of Testosterone in the Male Climacteric," *Journal of Clinical Endocrinology* 2 (1942): 238.

64. Paul de Kruif, *The Male Hormone* (New York: Harcourt, Brace, 1945).

65. De Kruif, *The Male Hormone,* 74.

66. De Kruif, *The Male Hormone,* 113.

67. De Kruif, *The Male Hormone,* 139.

68. De Kruif, *The Male Hormone,* 81.

69. De Kruif, *The Male Hormone,* 224, 226.

70. Kathryn Knowlton, Allan Kenyon et al., "Comparative Study of Metabolic Effects of Estradiol Benzoate and Testosterone Propionate in Man," *Journal of Clinical Endocrinology* 2 (1942): 671–95.

71. Carl Heller and Gordon Myers, "The Male Climacteric, Its Symptomatology, Diagnosis and Treatment," *JAMA* 126 (1944): 472–77.

72. Editorial, "Climacteric in Aging Men," *JAMA* 118 (1942): 458–59.

73. James Hamilton, "Testis Therapy," 267, in *Glandular Physiology and Therapy, A Symposium,* 1942, prepared under the auspices of the Council on Pharmacy and Chemistry of the American Medical Association.

74. Report of the Council on Pharmacy and Chemistry, "The Present Status of Testosterone Propionate: Three Brands, Perandren, Oreton and Neo-Hombreol (Roche-Organon) Not Acceptable for N.N.R.," *JAMA* 112 (1939): 1949–51. Note, too, how some physicians, like George Corner, wanted to restrict the use of male hormone to the specialists: *Hormones in Human Reproduction,* 239. He was very nervous about the interventions, mostly because of the Brown-Séquard, Steinach history.

75. Hamilton, "Testis Therapy," 271.

76. Medical Research Division, Schering Corporation, "Male Sex Hormone Therapy: A Clinical Guide" (n.p., 1941), 8, 9, 25.

77. Schering Corporation, "Male Sex Hormone Therapy," 27.

78. The advertisements are drawn from the journal *Endocrinology*, volume 22–23 (1938), volume 30 (1942).

79. Russell Cecil, *A Textbook of Medicine by American Authors* (Philadelphia: W. B. Saunders, 1930), 1206–7.

80. Russell Cecil, *A Textbook of Medicine by American Authors* (Philadelphia: W. B. Saunders, 1940), 1359.

81. Cecil, *Textbook of Medicine*, 1940 ed., 1360.

82. Cecil, *Textbook of Medicine*, 1940 ed., 1359.

83. See Elmer Sevringhaus, *Endocrine Therapy in General Practice* (Chicago: The Year Book, 1938), 161–65.

84. American Urological Association, *History of Urology*, vol. 2 (Baltimore: Williams & Wilkins, 1933), ch. 3.

85. Cost information gathered from *Drug Topics Price Book*, 1943–1944 ed. (New York: Topics Publishing Co., 1943); *Drug Topics Red Book*, 1946–1947 ed. (New York: Topics Publishing Co., 1946); *Drug Topics Red Book*, 1947–1948 ed. (New York: Topics Publishing Co., 1947). See U.S. Department of Labor, Bureau of Labor Statistics, Washington, D.C., Consumer Price Index, ftp://ftp.bls.gov/pub/special.requests/cpi/cpiai.txt, for consumer price equivalence. The price of testosterone did decline in 1947–48 to $22.86 per 25 milligrams. Ciba was also a supplier of the drug, charging $7 in 1943–44, then matching the others' prices thereafter.

86. See urologicalchannel.com/HealthProfiler/healthpro-testosterone.shtml.

87. Joyce Tenover, "Effects of Testosterone Supplementation in the Aging Male," *Journal of Clinical Endocrinology and Metabolism* 75 (1992): 1097.

88. R. A. Anderson and J. Bancroft, "The Effects of Exogenous Testosterone on Sexuality and Mood of Normal Men," *Journal of Clinical Endocrinology and Metabolism* 75 (1992): 1503–7.

89. J. Lisa Tenover, "Replacement Doses of Testosterone," *JCEM* 83 (1998): 3439.

90. Tenover, "Effects of Testosterone," 1095–97.

91. Louise D. Palmer, "Testosterone Replacement for Aging," *Plain Dealer*, March 13, 2000, 1F; Douglas A. Levy, "Is There Male Menopause?," *USA Today*, May 13, 1993, 14D; "Move Over Viagra," *Arizona Republic*, May 25, 2000, E1.

92. See feel21.com.

SEVEN   The Price of Growth

1. M. Mencer Martin and Lawson Wilkins, "Pituitary Dwarfism: Diagnosis and Treatment," *Journal of Clinical Endocrinology and Metabolism* 18 (1958): 680, 692.

2. Harold Stuart, "Normal Growth and Development During Adolescence," *NEJM* 234 (1946): 666–72.

3. C. H. Li and Harold Papkoff, "Preparation and Properties of Growth Hormone from Human and Monkey Pituitary Glands," *Science* 124 (1956): 1293–94; M. S. Raben, "Preparation of Growth Hormone from Pituitaries of Man and Monkey," *Science* 125 (1957): 883–84; J. C. Beck et al., "Metabolic Effects on Human and Monkey Growth Hormone in Man," *Science* 125 (1987): 884–85.

4. Max Goldzieher and Joseph Goldzieher, *Endocrine Treatment in General Practice* (New York: Springer Publishing, 1953), 4–10.

5. Herbert S. Kupperman, *Human Endocrinology* (Philadelphia: F. A. Davis, 1963), 56.

6. M. S. Rabin, "Treatment of a Pituitary Dwarf with Human Growth Hormone," *Journal of Clinical Endocrinology and Metabolism* 18 (1958): 901–2.

7. Kupperman, *Human Endocrinology,* 126.

8. Carl Gemzell and Choh Hao Li, "Estimation of Growth Hormone Content in a Single Human Pituitary," *Journal of Clinical Endocrinology and Metabolism* 18 (1957): 149–56.

9. John D. Crawford and Lester F. Soyka, "Growth Hormone," *The Practitioner* (Special Number, Advances in Treatment) 195 (1965): 550–53.

10. Martin and Wilkins, "Pituitary Dwarfism," 690.

11. NIH Cost-Reimbursement Contract with Johns Hopkins University (PH43-63-5764, April 19, 1963) to carry out "fundamental investigative work on human pituitary growth hormones and for the promotion of clinical investigation," 1.

12. S. Douglas Frasier, "The Not-So-Good-Old Days," *Journal of Pediatrics* 131 (1997): S2.

13. William Latimer to F. L. Mills, August 5, 1964, enclosing a copy of the newsletter. National Pituitary Agency Files.

14. William Latimer to Robert Bates, February 8, 1965. National Pituitary Agency Files.

15. Minutes of National Institute of Arthritis and Metabolic Diseases, *Contract Review Committee Meeting* (May 10, 1965), 1.

16. Frasier, "Not-So-Good Old Days," S2–S3.

17. Gilbert August et al., "Growth Hormone Treatment in the United States," *Journal of Pediatrics* 116 (1990): 902.

18. Lester Soyka, "Treatment of Hypopituitarism with Human Growth Hormone," in *Human Pituitary Growth Hormone,* Report of the Fifty-fourth Ross Conference on Pediatric Research (Columbus, Oh.: Ross Laboratories, 1965), 45–48.

19. John Money and Ernesto Pollitt, "Studies in the Psychology of Dwarfism," *Journal of Pediatrics* 68 (1966): 381.

20. Diane Rotnem et al., "Personality Development in Children with Growth Hormone Deficiency," *Journal of Child Psychiatry* 16 (1977): 417.

21. Hans-Christoph Steinhausen and Nikolaus Stahnke, "Psychoendocrinological Studies in Dwarfed Children and Adolescents," *Archives of Disease in Childhood* 51 (1976): 782.

22. Diane Rotnem et al., "Psychological Sequelae of Relative 'Treatment Failure' for Children Receiving Growth Hormone Replacement," *Journal of Child Psychiatry* 18 (1979): 10.

23. Rotnem et al., "Personality Development," 123.

24. Alfred Bongiovanni, "Commentary," *Journal of Pediatrics* 89 (1976): 1010.

25. Paul Brown, "Human Growth Hormone Therapy and Creutzfeldt-Jakob Disease," *Pediatrics* 81 (1988): 85–87.

26. Thomas Koch et al., "Creutzfeldt-Jakob Disease in a Young Adult with Idiopathic Hypopituitarism," *NEJM* 313 (1985): 731–33.

27. Paul Brown et al., "Potential Epidemic of Creutzfeldt-Jakob Disease from Human Growth Hormone Therapy," *NEJM* 313 (1985): 728–31; Colin Norman, "Virus Scare Halts Hormone Research," *Science* 228 (1985): 1176.

28. Judith Fradkin et al., "Creutzfeldt-Jakob Disease in Pituitary Growth Hormone Recipients in the United States," *JAMA* 265 (1991): 883–84.

29. P. Adlard and M. A. Preece, "Safety of Pituitary Growth Hormone," *Lancet* 344 (1994): 612–13.

30. S. Douglas Frasier and Thomas Foley Jr., "Creutzfeld-Jakob Disease in Recipients of Pituitary Hormones," *Journal of Clinical Endocrinology and Metabolism* 78 (1994): 1277–79; National Institute of Diabetes and Digestive and Kidney Diseases, "Follow-Up Study of NHPP Growth Hormone Recipients," Web site document, December 1999, 1–2.

31. C. J. Gibbs, Jr., and D. C. Gajdusek, "Infection As the Etiology of Spongiform Encephalopathy," *Science* 165 (1969): 1023–25.

32. The argument here follows closely on the paper presented by Robert Owen, one of the plaintiff's litigators in Britain on behalf of those who received growth hormone: "The Human Growth Hormone Creutzfeld-Jakob Disease Litigation," *Medico-Legal Journal* 65 (1996): 46–64.

33. Anonymous, "Growth Hormone," *The Medical Letter* 26 (1984): 81.

34. Brown, "Human Growth Hormone Therapy," 87.

35. See the excellent coverage of this story by Justin Gillis, "20 Years Later, Stolen Gene Haunts a Biotech Pioneer," *Washington Post*, May 17, 1999, A1.

36. Ron Rosenfeld et al., "Diagnostic Controversy: The Diagnosis of Childhood Growth Hormone Deficiency Revisited," *Journal of Clinical Endocrinology and Metabolism* 80 (1995): 1533.

37. Rosenfeld, "Diagnostic Controversy," 1533–34.

38. Jerome Grunt and I. David Schwartz, "Growth, Short Stature, and the

Use of Growth Hormone: Considerations for the Practicing Pediatrician," *Current Problems in Pediatrics* 22 (1992): 404; S. Douglas Frasier and Barbara Lippe, "The Rational Use of Growth Hormone During Childhood," *Journal of Clinical Endocrinology and Metabolism* 71 (1990): 269.

39. E. Kirk Neely and Ron Rosenfeld, "Use and Abuse of Growth Hormone," *Annual Reviews of Medicine* 45 (1994): 410.

40. Jens Jorgensen, "Human Growth Hormone Replacement Therapy," *Endocrine Reviews* 12 (1991): 190–91.

41. Z. Zadik et al., "Effect of Long-Term Growth Hormone Therapy on Bone Age and Pubertal Maturation in Boys with and Without Classical Growth Hormone Deficiency," *Journal of Pediatrics* 125 (1994): 189–95.

42. John Lantos, Mark Siegler et al., "Ethical Issues in Growth Hormone Therapy," *JAMA* 261 (1989): 1020–24.

43. Peter Hindmarsh and Charles Brook, "Final Height of Short Normal Children Treated with Growth Hormone," *Lancet* 348 (1996): 16.

44. J. M. Walker et al., "Treatment of Short Normal Children with Growth Hormone—a Cautionary Tale?," *Lancet* 336 (1990): 1331–34.

45. Erwin Bischofberger and Gunnar Dahlstrom, "Ethical Aspects of Growth Hormone Therapy," *Acta Paediatrica Scandinavica*, supplement 362 (1989): 16.

46. Douglas S. Diekema, "Is Taller Really Better? Growth Hormone Therapy in Short Children," *Perspectives in Biology and Medicine* 34 (1990): 116.

47. Lantos et al., "Ethical Issues," 1020, 1023.

48. Louis Underwood and Patricia Rieser, "Is It Ethical to Treat Healthy Short Children with Growth Hormone," *Acta Paediatrica Scandinavica*, supplement 362 (1989): 19; David Allen and Normal Fost, "Growth Hormone Therapy for Short Stature: Panacea or Pandora's Box?," *Journal of Pediatrics* 117 (1990): 17; Grunt and Schwartz, "Growth, Short Stature," 408.

49. Louis Underwood, "Growth Hormone Therapy for Short Stature," *Hospital Practice* 27 (1992): 192.

50. Allan and Fost, "Growth Hormone Therapy for Short Stature," 20.

51. Underwood and Rieser, "Is It Ethical to Treat," 19.

52. Hindmarsh and Brook, "Final Height of Short Normal Children," 13.

53. Frasier and Lippe, "The Rational Use of Growth Hormone," 272; Karl C. Moore et al., "Clinical Diagnosis of Children with Extremely Short Stature and Their Response to Growth Hormone," *Journal of Pediatrics* 122 (1993): 691.

54. In October 1991, the University of Wisconsin sponsored a forum on the pros and cons of growth hormone treatment, which was published as "Access to Treatment with Human Growth Hormone: Medical, Ethical, and Social Issues," in *Growth, Genetics, and Hormones* 8 (1992), supplement 1. See pp. 37, 68, for quotations.

55. "Access to Treatment," 48.

56. "Access to Treatment," 60.

57. "Access to Treatment," 71.

58. "Report of the NIH Human Growth Hormone Protocol Review Committee" (October 2, 1992). Thanks to Carol Tauer for sharing it with us; she obtained it through the Freedom of Information Act. The report was submitted to Bernadette Healy, director of the NIH, from co-chairs William Friedwald, medical director of Metropolitan Life Insurance Company, and Loretta Kopelman, chair of the Department of Medical Humanities, East Carolina University School of Medicine.

59. See the fascinating analysis of this story and the links to genetic enhancement by Carol A. Tauer, "The NIH Trials of Growth Hormone for Short Stature," *IRB* 16 (1994): 1–9. Tauer explicitly notes that by virtue of this decision, the panel and the NIH "open the door" for enhancement research, both on adults and children (p. 8).

60. "Charity Tactic by Genentech Stirs Questions," *Wall Street Journal*, August 12, 1994, B1, B4.

61. Statement of Fran Price, executive director, HGH, before the Subcommittee of Regulation, Business Opportunities, and Technology, U.S. House of Representatives, Small Business Committee, October 12, 1994.

62. "Written Statement of Genentech," Committee on Small Business, U.S. House of Representatives, October 12, 1994.

63. U.S. District Court of Minnesota, Fourth Division, Case Number Cr. 4-94-95, *United States of America vs. David R. Brown et al.* On file in the district court, August 8, 1995 (second day of trial).

64. *United States of America vs. David R. Brown et al.*, II-A-1.

65. *United States of America vs. David R. Brown et al.*, II-A-126.

66. Maura Lerner, *Minneapolis Star Tribune*, "Doctor's Conviction Put Aside Due to Juror Misconduct," January 15, 1996; State Court of Minnesota, Court of Appeals, C2-97-817: *D.A.B. et al. vs. David R. Brown et al.*, November 4, 1997.

67. M. S. Raben, "Clinical Use of Human Growth Hormone," *NEJM* 266 (1962): 82–86.

68. Franco Salomon et al., "The Effects of Treatment with Recombinant Human Growth Hormone on Body Composition and Metabolism in Adults with Growth Hormone Deficiency," *NEJM* 321 (1989): 1797–1803; J. O. Jorgensen et al., "Beneficial Effects of Growth Hormone Treatment in GH-Deficient Adults," *Lancet* 1 (1989): 1221–25.

69. Daniel Rudman, "Effects of Growth Hormone in Men over 60 Years Old," *NEJM* 323 (1990): 1–6.

70. S. E. Inzucchi and R. J. Robbins, "Growth Hormone and the Maintenance of Adult Bone Mineral Density," *Clinical Endocrinology* 45 (1996): 665–73; Robert Marcus, "Skeletal Effects of Growth Hormone and IGF-1 in Adults," *Hormone Research* 48 (1997): 60–64.

71. Janet Vittone et al., "Effects of Single Nightly Injections of Growth

Hormone-Releasing Hormone (GHRH 1-29) in Healthy Elderly Men," *Metabolism* 46 (1997): 89–96.

72. Nina Vahl et al., "Abdominal Adiposity and Physical Fitness Are Major Determinants of the Age Associated Decline in Stimulated GH Secretion in Healthy Adults," *Journal of Clinical Endocrinology and Metabolism* 8 (1996): 2213; S. Boonen, J. Aerssens et al., "Age-Related Endocrine Deficiencies and Fractures of the Proximal Femur—Implications of Growth Hormone Deficiency in the Elderly," *Journal of Endocrinology* 149 (1996): 10.

73. Maxine Papadakis et al., "Growth Hormone Replacement in Healthy Older Men Improves Body Composition but Not Functional Ability," *Annals of Internal Medicine* 124 (1996): 715.

74. J. J. Chipman et al., "Safety Profile of GH Replacement Therapy in Adults," *Clinical Endocrinology* 46 (1997): 473–81.

75. Renee Page et al., "An Account of the Quality of Life of Patients After Treatment for Non-Functioning Pituitary Tumors," *Clinical Endocrinology* 46 (1997): 404. Or as yet another team concluded: "Many questions related to dosage, benefit, safety and tolerance need to be critically addressed before GH can be considered for the somatopause." Ken Ho and D. M. Hoffman, "Aging and Growth Hormone," *Hormone Research* 40 (1993): 85.

76. Mary Lee Vance and Nelly Mauras, "Growth Hormone Therapy in Adults and Children," *NEJM* 341 (1999): 1210; Jayaraj Kandaswamy, "Should Doctors Give Hormones to Healthy Elders?," *North Carolina Medical Journal* 60 (1999): 340–45.

77. John Morley, "Growth Hormone: Fountain of Youth or Death Hormone?," *Journal of the American Genetics Society* 47 (1999): 1475–76.

78. National Institute on Aging, "Pills, Patches, and Shots: Can Hormones Prevent Aging?," October 6, 1999, release on its Web site, nih.gov/nia.

79. D. G. Johnson, "Growth Hormone Deficiency and Quality of Life in Hypopituitary Adults," *Clinical Endocrinology* 46 (1997): 407–8; Horace M. Perry III, "The Endocrinology of Aging," *Clinical Chemistry* 45 (1999): 1369–76.

80. Paul Carroll, Emanuel Christ, and the members of the Growth Hormone Research Society Scientific Committee, "Growth Hormone Deficiency in Adulthood and the Effects of Growth Hormone Replacement: A Review," *Journal of Clinical Endocrinology & Metabolism* 83 (1998): 383–95. See also Gudmundar Johannsson et al., "Growth Hormone Treatment of Abdominally Obese Men Reduces Abdominal Fat Mass, Improves Glucose Lipoprotein Metabolism, and Reduces Diastolic Blood Pressure," *Journal of Clinical Endocrinology & Metabolism* 82 (1997): 727–34.

81. Jukka Takala et al., "Increased Mortality Associated with Growth Hormone Treatment in Critically Ill Adults," *NEJM* 341 (1999): 785–92, 837–39. Note trauma and burn exclusion.

82. Karen T. Coschigano et al., "Assessment of Growth Parameters and Life Span in GHR/BP Gene-Disrupted Mice," *Endocrinology* 141 (2000): 2612.

83. Editorial, "Not Quite the Fountain of Youth," *New York Times*, July 6, 1990, A24; Natalie Angier, "Human Growth Hormone Reverses Effects of Aging," *New York Times*, July 5, 1990, A1; Geoffrey Cowley and Mary Hager, "Can Hormones Stop the Clock?," *Newsweek*, July 16, 1990, 66.

84. AP release of July 5, 1990.

85. Richard Saltus, "Tinkering with the Mechanisms of Aging," *Boston Globe*, August 6, 1990, 25.

86. Kim Painter, "Elderly See Benefits from Growth Hormone," *USA Today*, April 15, 1996, 1D; "Growth Hormone Fails to Reverse Effects of Aging, Researchers Say," *New York Times*, April 14, 1996, 608; Robert Cooke, "Fountain-of-Youth Springs a Leak," *New York Newsday*, April 16, 1996, A52.

87. Maureen West, "The Youth Pill," *Arizona Republic*, May 3, 1998, A1; Michele Lesie, "The Hormone of Youth," *Plain Dealer*, Sunday Magazine, October 18, 1998; Thomas Maier, "Hormones a Growing Concern," *New York Newsday*, November 17, 1998, C8.

88. American Academy of Anti-Aging Medicine, "Frequently Asked Questions: What Types of Doctors Practice Anti-Aging Medicine." See worldhealth.net.

89. Ann Oldenberg, "Boomers Believe They've Found a Fountain of Youth in a Syringe," *USA Today*, November 14, 2000, 1A.

90. Alex Kuczynski, "Fountain of Youth," *New York Times*, April 19, 1998.

91. Thomas Maier, "Hormone a Growing Concern," *New York Newsday*, November 17, 1998, C8.

92. See the Cenegenics Web site, 888younger.com, and the Web site of the Anti Aging Medical Associates of Manhattan, antiagingny.com.

93. Deborah Shelton, "Dipping into the Fountain of Youth," *American Medical News*, December 4, 2000.

94. For an update survey, see Gina Kolata, "Chasing Youth, Many Gamble on Hormones," *New York Times*, December 22, 2002.

95. Robert Langreth, "Sweet Syringe of Youth," *Forbes*, December 11, 2000.

96. Lesie, "The Hormone of Youth."

EIGHT   Peak Performance

1. Tim Tully, "Discovery of Genes Involved with Learning and Memory: An Experimental Synthesis of Hirschian and Benzerian Perspectives," *Proceedings of the National Academy of Sciences of the United States of America* 93 (1996): 13,463.

2. Alan Gelperrin, "Flies, Genes, and Memory Engineering," *The Biological*

Bulletin 191 (1996): 139. See also Mark Mayford, Ted Abel, and Eric Kandel, "Transgenic Approaches to Cognition," *Current Opinion in Neurobiology* 5 (1995): 141–48.

3. Gelperin, "Flies, Genes," 140.

4. Jim Dezazzo and Tim Tully, "Dissection of Memory Formation: From Behavioral Pharmacology to Molecular Genetics," *Trends in Neurosciences* 18 (1995): 212–18.

5. C. Chen and S. Tonegawa, "Molecular Genetic Analysis of Synaptic Plasticity, Activity-Dependent Neural Development, Learning and Memory in the Mammalian Brain," *Annual Review of Neuroscience* 20 (1997): 158–59.

6. M. A. Wilson and S. Tonegawa, "Synaptic Plasticity, Place Cells and Spatial Memory: Study with Second Generation Knockouts," *Trends in Neurosciences* 20 (1997): 102.

7. Not only genetic manipulation but cellular substances and pharmacological agents are being applied to laboratory mice to enhance their memory. One favorite is GM1, a substance (technically a ganglioside containing glycosphingolipids) that is present at higher concentrations in the nervous system than in any other body tissue. GM1 is only one of many substances that seem to improve memory in laboratory animal experiments. Psychopharmacology journals, for example, report that the administration of kynurenic acid (KYNA), a substance found in the mammalian brain, also increases learning and memory in rats; so too, compounds like milacemide (an anti-epileptic drug), which increase the amount of glycine, increase neural receptor activity, and improve learning and memory. Evidence, however, has been accumulating that milacemide may be toxic to the liver. A variant of KYNA, milademide, has also been reported to enhance memory in normal and amnesiac humans, and choline, a dietary component essential for the normal functioning of all cells (eggs are rich in it), may have the same effects as well. Chen and Tonegawa, "Molecular Genetic Analysis," 178.

8. F. P. Zemlan et al., "Double-Blind Placebo-Controlled Study of Velnacrine in Alzheimer's Disease," *Life Sciences* 58 (1996): 1823–32.

9. S. C. Samuels and K. L. Davis, "A Risk-Benefit Assessment of Tacrine in the Treatment of Alzheimer's Disease," *Drug Safety* 16 (1997): 66–77.

10. Richard Mayeux and Mary Sano, "Drug Therapy: Treatment of Alzheimer's Disease," *NEJM* 341 (1999): 1670–79.

11. Sander Gilman, "Alzheimer's Disease," *Perspectives in Biology and Medicine* 40 (1997): 230–45.

12. Rudolph Tanzi and Ann Parson, *Decoding Darkness* (Cambridge, Mass.: Perseus Publishing, 2000).

13. Gilman, "Alzheimer's Disease," 237.

14. Lindsay Farrer et al., "Effects of Age, Sex, and Ethnicity on the Association Between Apolipoprotein E Genotype and Alzheimer Disease," *JAMA* 278 (1997): 1349–56.

15. Mayeux and Sano, "Treatment of Alzheimer's Disease," 1670–79.

16. M. W. Bondi et al., "Neuropsychological Function and Apolipoprotein E Genotype in the Preclinical Detection of Alzheimer's Disease," *Psychology and Aging* 14 (1999): 295–303; B. J. Small et al., "Is APOE-e4 a Risk Factor for Cognitive Impairment in Normal Aging?," *Neurology* 54 (2000): 2082–88.

17. David Rowe, *The Limits of Family Influence: Genes, Experience, and Behavior* (New York: Guilford Press, 1994), 63.

18. Rowe, *The Limits of Family Influence*, 64.

19. Kenneth Schaffner, "Complexity and Research Strategies in Behavioral Genetics," in *Behavioral Genetics: The Clash of Culture and Biology*, ed. Ronald Carson and Mark Rothstein (Baltimore: Johns Hopkins University Press, 1999), 68.

20. Jonathan Benjamin, Lin Li et al., "Population and Familial Association Between the D4 Dopamine Receptor Gene and Measures of Novelty Seeking," *Nature Genetics* 12 (1996): 83.

21. Joseph Alper and Jonathan Beckwith, "Genetic Fatalism and Social Policy: The Implications of Behavior Genetics Research," *Yale Journal of Biology and Medicine* 66 (1993): 511–24.

22. Paul Billings, Jonathan Beckwith, and Joseph S. Alper, "The Genetic Analysis of Human Behavior: A New Era?," *Social Science and Medicine* 35 (1992): 227–38. See also Schaffner, "Complexity and Research Strategies," 62.

23. Gregory Carey and David DiLalla, "Personality and Psychopathology: Genetic Perspectives," *Journal of Abnormal Psychology* 103 (1994): 36.

24. Ming Tsuang, "Schizophrenia: Genes and Environment," *Biological Psychiatry* 47 (2000): 210.

25. Preben Bo Mortensen, Carsten Bocker Pedersen, Tine Westergaard et al., "Effects of Family History and Place and Season of Birth on the Risk of Schizophrenia," *NEJM* 340 (1999): 603–8. See also Rowe, *The Limits of Family Influence*, 17; Ming Tsuang, W. S. Stone, and S. V. Faraone, "Schizophrenia: A Review of Genetic Studies," *Harvard Review of Psychiatry* 7 (1999): 185–207; Ming Tsuang, "Schizophrenia: Genes and Environment," 210–20; S. O. Moldin and I. I. Gottesman, "At Issue: Genes, Experience, and Chance in Schizophrenia," *Schizophrenia Bulletin* 23 (1997): 547–61.

26. Alan Sanders and Pablo Gejman, "Influential Ideas and Experimental Progress in Schizophrenia Genetics Research," *JAMA* 285 (2001): 2831–33.

27. Ann Pulver, "Search for Schizophrenia Susceptibility Genes," *Biological Psychiatry* 47 (2000): 227.

28. Moldin and Gottesman, "At Issue," 554.

29. John Crabbe, "Use of Genetic Analyses to Refine Phenotypes Related to Alcohol Tolerance and Dependence," *Alcoholism: Clinical and Experimental Research* 25 (2001): 288–92. See also Mark Schuckit, "New Findings in the Genetics of Alcoholism," *JAMA* 281 (1999): 1875–76.

30. Paul Buckland, "Genetic Association Studies of Alcoholism—Problems with the Candidate Gene Approach," *Alcohol & Alcoholism* 36 (2001): 99–103.

31. Daniel Lieberman, "Children of Alcoholics: An Update," *Current Opinion in Pediatrics* 12 (2000): 336–40.

32. M. M. Vanyukov and R. E. Tarter, "Genetic Studies of Substance Abuse," *Drug and Alcohol Dependence* 59 (2000): 101–33.

33. Kenneth Blum, Ernest Noble et al., "Allelic Association of Human Dopamine D2 Receptor Gene in Alcoholism," *JAMA* 263 (1990): 2055–60. See also Kenneth Blum and Ernest Noble, "The Sobering D2 Story," *Science* 265 (1994): 1347.

34. Elliot Gershon, quoted in Constance Holden, "A Cautionary Genetic Tale: The Sobering Story of D2," *Science* 264 (1994): 1696.

35. Ernest Noble, "Addiction and Its Reward Process Through Polymorphisms of the D2 Dopamine Receptor Gene: A Review," *European Psychiatry* 15 (2000): 79–89.

36. Noble, "Addition and Its Reward Process," 87.

37. Karel Capek as quoted in Richard Kane, "The Defeat of Aging Versus the Importance of Death," *Journal of the American Gerontological Society* 44 (1996): 322.

38. Leonard Hayflick, "How and Why We Age," *Experimental Gerontology* 33 (1998): 652.

39. Robert Reis and Robert Ebert II, "Genetics of Aging Current Animal Models," *Experimental Gerontology* 31 (1996): 69–81.

40. Michael Rose and Theodore Nusbaum, "Prospects for Postponing Human Aging," *Federation of American Societies for Experimental Biology Journal* 8 (1994): 925.

41. Rose and Nusbaum, "Prospects for Postponing Human Aging," 925.

42. Rose and Nusbaum, "Prospects for Postponing Human Aging," 926.

43. The most accessible discussion of Hayflick and the Hayflick limit is Stephen S. Hall, *Merchants of Immortality* (Boston: Houghton Mifflin, 2003).

44. Andrea Bodnar, Michel Ouellette et al., "Extension of Life-Span by Introduction of Telomerase into Normal Human Cells," *Science* 279 (1998): 351–52.

45. Michael Fossel, "Telomerase and the Aging Cell: Implications for Human Health," *JAMA* 275 (1998): 1732.

46. Fossel, "Telomerase and the Aging Cell," 1733.

47. Fossel, "Telomerase and the Aging Cell," 1734.

48. Carmela Morales, Shawn Holt et al., "Absence of Cancer-Associated Changes in Human Fibroblasts Immortalized with Telomerase," *Nature Genetics* 21 (1999): 115.

49. Morales, "Absence of Cancer-Associated Changes," 116.

50. Jim Wang, Gregory Hannon et al., "Risky Immortalization by Telomerase," *Nature* 405 (2000): 755.

51. Wang, "Risky Immortalization," 755.

52. James D. Watson, *DNA: The Secret of Life* (New York: Alfred A. Knopf, 2003), 398–400.

53. Margaret Atwood, *Oryx and Crake* (New York: Doubleday, 2003), 22, 55. Appearing within this period, and taking a less apocalyptic approach toward anti-aging enhancements, is the novel by Michael Byers, *Long for This World* (Boston: Houghton Mifflin, 2003).

54. Atwood, *Oryx and Crake*, 57.

55. Atwood, *Oryx and Crake*, 248.

56. Carl Elliott, *Better Than Well* (New York, W. W. Norton, 2003), 6–7.

57. Elliott, *Better Than Well*, 165.

58. Elliott, *Better Than Well*, 207. Readers will also be interested in the recent study by Stephen S. Hall, *Merchants of Immortality* (Boston: Houghton Mifflin, 2003). This book, like the others referenced immediately above, apeared too late to structure our own arguments, but we found ourselves in close agreement with many of the positions Hall adopts, especially in his Epilogue.

Acknowledgments

WE RECEIVED EXEMPLARY ASSISTANCE in the course of researching and writing this book and are delighted to acknowledge it. Our major source of support was the ELSI program of the Human Genome Institute of the National Institutes of Health (grant RO1 HG 01505), and Elizabeth Thomson was especially helpful and encouraging. We received supplementary funding from the Samuel and May Rudin Foundation, with Jack Rudin once again facilitating our work.

We took regular counsel from colleagues at Columbia, most notably Isidore Edelman and Jennifer Bell; in addition, Harold Edgar, David Silvers, and Georgiana Jagiello reviewed the final manuscript. Needless to say, they have no responsibility for errors or misinterpretations. Conversations with, among others, Eric Kandel, Norman Hugo, Irene Sills, and Yves Illouz were particularly illuminating. We had the opportunity to present our findings in a variety of forums, including the ELSI Tenth Anniversary Conference ("Human Gene Transfer Beyond Life Threatening Disease"), the Museum of Natural History, and the New York Academy of Medicine. These sessions, as well as those at other medical schools, helped us clarify our arguments.

A number of medical students assisted us over the years. We are grateful to Vlad Manual, Ajay Kirtane, Alex Haynes, Allegra Broft, Catherine Fullerton, Katie Kraft, and Andrew Bomback. The staff at the Center for the Study of Society and Medicine was efficient in tracking down citations and sources and invaluable in the preparation of the manuscript. Our deep thanks to Beth Matish, Tiffany Rounsville, Irina Vodones, and Martin Rivlin. We received many courtesies from librarians and archivists and were especially dependent on the staff at the Hammer Library at Columbia P&S, the New York Academy of Medicine, the Bancroft Library at the University of California, Berkeley (the Gertrude Atherton papers), and the Library of Congress.

We had the good fortune to complete and revise drafts of the book at two exceptional settings. We were resident scholars at the Rockefeller Foundation's Bellagio Conference Center and, for a second time, were privileged to enjoy the gracious hospitality of Gianna Celli; that our stay came in the immediate aftermath of 9/11 made it all the more significant. Several months later, we were guests at the Santa Fe Institute. The chance to discuss enhancement issues with a diverse group of computational biologists, economists, and mathematicians was as challenging as it was stimulating.

We were encouraged and facilitated in our work by our agent and friend, Collin Stanley (of Janklow Morris). Our editors at Pantheon, Dan Frank and Rahel Lerner, were unfailingly supportive, and even more important, gave us sound advice on how to better organize and present our material. We were lucky to be able to rely on such a trio.

Index